Claude ROUQUETTE

LOW BIOLOGICAL ENERGIES

Claude ROUQUETTE

LOW BIOLOGICAL ENERGIES

Essay on Biological Evolution and the Transformations of Civilization

ScienciaScripts

Contents

The naturalist suite on complex processes in Biological Evolution and the Transformations of Civilization consists of four chronological volumes:

Volume I: Les fougeres noires.

Volume II: L'Euprocte des Pyrenees.

Volume III: The Cevennes beaver.

Volume IV: Indri indri, a journey to the origins of humanity.

Followed by three essays:

-       Low biological energies.

-       L'Homme desarticule.

-       The seventh string.

Other works by the author on the history of the navy and the Cevennes, an introduction to scientific exploration of the seas, oceans and continents.

-       Le college de Neptune. (Les Presses du Midi)

-       Memoires de mer, cevenoles.

Preface

After writing several books on complex evolutionary processes, Claude Rouquette's essay reveals the maturity of an in-depth reflection on the involution of living organisms and the singular consëquences that result from our social interactions and cohësions. In his Darwinian research, which combines his methodical knowledge of evolutionary biology with a civilisational approach, his experience of extreme environments and crisis situations has given him the intellectual space and scientific rationality to develop his gifts as an analyst and humanist.

Today, after an exceptional career as a marine historian and naturalist, he has found in the 'elements of divergent transition' the material for an original essay on the evolution of species whose trajectories appear to be determined but are nonetheless linked to discreetly random fluctuations.

I knew the author well at a time of great operational instability and dramatic change. Our long discussions, on missions in the Middle East and on board aircraft carriers equipped with nuclear weapons at the time, left a deep impression on me and made me realise that, in addition to being a recognised security specialist, he was a man with a remarkable desire to learn and understand, always ready to get involved when knowledge was inadequate.

It was also a time when personal commitment and constructive innovation could flourish in a spirit of responsibility in the pursuit of the general interest, freeing themselves, when necessary, from castrating rules.

In his work, with the resolution, patience and meticulousness for which he is renowned, he describes the most subtle and complex components of the understanding of being. Etude's originality lies in identifying the weak signals of what we commonly call alea and chance, in order to make Elucidation of the most ordinary or unusual event probable. This is the principle of reason and knowledge of the inner workings of a mind that is developed by conscience in the midst of sublime instincts.

It makes it possible to understand even the most seemingly innocuous mutations. Through observation of reality, guided by a systemic logic, he reveals the episodes that depend on this or that criterion of biological life, making lucid reference to previous research on the subject. He reminds us of them with precision and references.

The first quality of this work is that it presents situations as potential factors for characterising the physical, biological and social systems of complex processes, encouraging reflection on the dynamics of the endogenous and exogenous vectors of variability in the criteria for the contingent evolution of species.

The contribution made here to a better understanding of low biological energies gives this collection a force behind our emotions, reminding us with humility of the unfinished transition of our human condition.

Captain (h)
Jean-Claude RICHARD
Former Director General of the Fondation mediterraneenne
d'etudes strategiques
Le Beausset, 15 April 2021

Acknowledgements

I would like to express my gratitude to Rear-Admiral (2s) Jacques Branellec for the trust he placed in me during my postings to the aircraft carrier *Foch* and the *Marseille Marine Fire Brigade*.

I would like to express my sincere thanks to Captain Jean-Claude Richard, who kindly agreed to preface this essay. His career as a naval officer and his path in civilian society are an example and an encouragement for young generations looking to the future.

I can't forget Major Jean-Jacques Desroziers, my fellow officers and naval officers, safety mechanics, carpenters and firemen with whom I had the honour of serving in the French Navy.

Many thanks for their friendship to naturalists Sylain Mahuzier and Jean-Pierre Sylvestre.

This essay is a tribute to the late Doctor of Biology Chomin Cunchillos, author of *"Pathways to Emergence"*. As assistant to the Spanish biologist Faustino Cordon (1909-1999) and his daughter Teresa, he worked on the theory of integration level units.

Thank you, to Patrick Tort, philosopher and historian of science, Director of the Charles Darwin Institute International.

Foreword

During the writing of the four-volume naturalist suite, as part of my fundamental and applied research into ёbiological evolution and the transformations of civilisation, a number of questions arose about the energies *'at play'* in the life cycles of the plant and animal species observed. Complex evolutionary processes are by definition highly integrated, their combinatorial nature generating fundamentally random variations in their Darwinian relationship to selective constraints and random circumstances.

During the revolution of a species, from its origins onwards, because of this repeated evolutionary contingency, during the reproduction of successive generations, the stochasticity that results from the phase transitions between linearity and non-linearity during singular divergences can be observed on the reconstituted phylogenetic trees of the species. During these bifurcations, we can analyse their evolutionary trajectories at different levels of integration, which can be broken down into strata representing their biological units. At each divergence, the critical stochasticity values that cause these singularities are normalised dimensionless adaptativeness coefficients, with which it is possible to explore the remarkable evolutionary patterns of a species, such as its speciation, diversity or extinction. The method of analysing complex evolutionary processes provides access to the different units of integration level, using the appropriate scientific disciplines to identify the most indistinguishable vital coherences.

This naturalistic essay on low biological energies takes us to the frontier between quantum mechanics and statistical physics.

From the atom to man.

As a prelude to this naturalist dissertation on low biological energy, I relate my experience as a naval officer specialising in safety and the environment to explain the motivation that led me to study the complex processes of biological evolution and the transformations of civilisation over a period of some forty years.

At the end of this preliminary account, I set out the fundamental principles that I had initially retained to constitute my first modëles for solving complex ëvolutive processes. In order to argue my proposals, I have rësumë the main lessons drawn from fundamental and applied ëstudies in a chronological naturalistic suite rëdigëe in four volumes, in which I have progressively introduced key concepts.

In this way, the stochastic phase transition between linearity and non-linearity has enabled me to probe the singular spaces of divergence and their associated strata, which *'reflect'* the adaptativeness that manifests itself during the revolution of species. The trajectories of their phylogenies, patiently and sparingly reconstructed by paleontologists and systematists, present numerous highly divergent tree-like ramifications, the dynamics of which can be likened to mosaics resulting from incessant complex interactions.

The tome devoted to "***Castor** fiber, des Cevennes"* made me aware of the body's reactions to the thermal and hydric stress suffered by a small population of përiodically isolated beavers during summer droughts on a tributary of the Cëze. During twenty years of observations, I have ëtudiedë a probable case of speciation by developing approach models derived from Castorid phylogeny, corrected by test models by assimilating data from observations, to determine a probability of divergence of this small population subject to severe natural and anthropogenic selective constraints.

In the volume devoted to the Pyrenean skink, ***Calotriton** asper asper*, this Urodele asked me many questions about the adaptive capacities of Lissamphibians, and in particular about their living conditions at low temperatures at altitude:

- The adaptation of its cutaneous respiration to life in water and on land at altitude and at low temperatures, particularly during hibernation.
- Its development and metamorphosis.
- Its ability to regenerate an organ.

So, from the first to the fourth volume of the naturalist suite, going back to the fossilised plants of the Carboniferous Cevennes to the Lemurian, ***Indri** indri*, in a journey to the origins of Humanity: I came across the problem of managing the energy that contributes to the adaptability of a species, to maintain its vital coherence, during its relative perennity in space and time, in the course of its biological evolution.

As far as the human species is concerned, in view of its ëvolution and history, the socio-economic use of resources as *"matter and energy"* necessary for the transformations of civilisation seemed to me to be essential to ensure the indispensable vital coherences by generating the social cohesions that maintain society in its civilised state, a reversive effect (Tort,1983), exclusive, among living beings.

Compared with the low biological energies used within a cell, the power level of our energy needs is exorbitant, if not excessive. To acquire this knowledge, I have seen whole swathes of civilisation crumble during my travels around the world, coming into contact with populations in the throes of incessant crises and deadly conflicts in which I was involved to help secure oil supplies, a natural resource wasted on the highways of over-consumption and leisure that pretends to be green...

As my research progressed, there were levels of resolution that posed transitional problems that I encountered both in evolutionary biology, and in the civilisation weakened by our most deleterious actions, for the future of humanity, species and their environments. I had to explore the keys to life from the atomic level upwards, using quantum mechanics whose effects are, to say the least, discrete and random, compared with the statistical physics of the thermo-hydro-dynamic domains, which are more accessible to our senses...

In the first chapter, we present two not-so-elemental particles that are very much involved in physiology, catabolism and cellular metabolism. The proton and the electron are described by quantum mechanics and the field theory of the same name.

At the end of this essential presentation, we will deal with this aspect of matter-energy, as a particle and a wave, whose fields are manifested in an organelle of the animal cell, the mitochondrion.

In the second chapter, I will take time to describe the mitochondria, in particular the transport of electrons in the basic protëine complexes embedded in the inner membrane of this organelle of the animal cell, together with the translocation of protons between the matrix and the inter-membrane space, where a threshold pulse proton field is formed, which we are trying to model.

In the third chapter, we will discuss the capacity of this proton field to coordinate cellular and organismal activity, to 'see' how such a fundamentally quantum process emerges and becomes singularised at each level of biological resolution.

By way of conclusion, after formulating a few hypothetical avenues of research into low biological energies and integration level units applied to the theory of complex evolutionary processes. We will dwell on the energy-generating strata of civilisation, which are currently generating highly worrying situations that require long-term scientific exploration, both globally and locally.

I've experienced these critical situations during successive tours of duty aboard French Navy combat ships in overseas operations and in France.

During my years of service in the Dëfense Nationale, I wore on the sleeve of my petty officer's jacket and on my naval officer's jacket, from second petty officer to major, the insignia for the speciality of security electro-mechanic (E.m.sec.). Around the 'S' for security, surrounded by a jagged wheel and a few lightning bolts, were two electrons that gave us expertise in defence against nuclear aggressors and other inventions of mass destruction, whether bacteriological or chemical.

This speciality was created in the 1960s, based on the experience of mechanics, electricians and marine carpenters who fought fire and waterways during naval battles on board Royal Navy vessels. The E.m.sec. and the marine fire-fighters are heirs to a long experience of safety organisation on board the Navy's combat ships and the lessons learned from conflicts, from the last world wars to the most recent armed conflicts (Vietnam, Korea, Falklands, Iraq, Balkans, Afghanistan, Syria, etc.).

After being posted as a seaman mechanic aboard the aviso-escorteur 'Commandant Bory' on a campaign in the Indian Ocean in 1969, I took part in one of the first courses in this speciality in the company of a few choufs, quartermasters with experience of far-flung campaigns. Companions of the sea with whom we learned to fight against adversity at the French Navy's Ecole des Marins Electriciens et de Securite (E.M.E.S.), based in Querqueville, near

Cherbourg.

1. Maritime training booklet
for Quartermaster E.m.sec.

Our company was summarily housed in old barracks, in which we had classrooms and partials for our technological training and combat damage training, for six months, in order to obtain a technical aptitude certificate. We resumed the ritual of military and maritime training, marching in step and uniform inspections, with a steady rhythm of theoretical courses alternating with the practical fire-fighting and waterway exercises required by our emlbirque spe;ciali^. Near our billet, on the seafront, an old hull was moored to train us to fight against combat damage, fire, water, bacteriological, chemical and nuclear aggressors. Nuclear tests were being carried out in the Pacific to contribute to the development of deterrence in the naval, air and land forces of the French national defence. This experience was to have a major influence on my later career...

2. The "*Lucifer*" at low tide

The Royal Navy frigate "*Windrush*", cëdëe to the Free French Naval Forces, became "*La Decouverte*" (1943) and took part in the Normandy dëbarquement, dësarmëe in 1959, She was used for safety training of the fleet's crews under the name of "*Lucifer I, then Lucifer II*" after a number of improvements to the training modules for fighting fires, waterways and aggressive, nuclear, biological and chemical (*NBC) agents*. For daytime training and night-time cëXëHлë drills, we had premises that were equipped to deal with various types of disaster, such as those in boiler rooms and under machine floors, those of the deep-fat fryers in the galleys, those of the very coni'iiK' bar room, outside, those litë on gun mountings, the waterways dëroulaient in a bilge...

3. Putting out a hydrocarbon tank fire

Equipped with our parkas and self-contained breathing apparatus we pënëtaited in very cold water, trying to seal off the waterways. Each ëquad valiantly pressed pinoches or plugs into the bores to limit the ingress of icy water, which the pressure, regulated by the experienced instructors, pushed back and ejected, despite our efforts deployed in the smoke and exploding fartards. Stubbornly, and with the help of our solidarity, we tried to contain the flood which invaded the compartment, threatening the buoyancy and stability of the ship.

It was in this way that I began to understand the states of criticality and unstable equilibrium that I would find in the transition between the linear and non-linear nature of complex evolutionary processes.

At the end of the course, I boarded the *"Orage"* landing barge, which provided logistical support for the nuclear explosions in the Pacific, and saw these monstrous mushrooms rise above the Mururoa atoll, then witnessed the underground explosions.

4. Over-critical, divergent nuclear explosion.

After two years of intense sailing in the Pacific Islands, I returned to the Centre d'Instruction Naval de Querqueville for the E.m.sec. brevet superieur course, a refresher course to become an experienced naval officer, which would take concrete form when I embarked on the wing supply tanker *"La Seine"*. It was an exciting assignment on this old hull, which supplied the aircraft carriers and frigates of the aeronaval group engaged in external operations in the Mediterranean, the Near and Middle East, at a time when the war was still cold and heating up dangerously in Lebanon and the Persian Gulf.

Without ^pit from premier-maitre to maitre-principal, I would embark successively on the aircraft carriers *"Clemenceau"* and *"Foch"* engagës sans interruption dans l'Ocëan Indien et le golfe Persique, toujours en ё^ИШси, notre mission ё-tait de protëger les routes du pëtrole, tres convokes.

5.   Rescue training in asbestos suits.

I would return to the École de Sëcuritë as an instructor in the Nuclear, Biological and Chemical section, after an interarmëes NBC course at the École des armes spëciales in Grenoble, I would take the opportunity to pass my radiation protection technician course. Along with my fellow students[1]My mission, along with my fellow students, was to develop an NBC defence course to train the crews of combat ships, as chemical weapons were making a dangerous comeback in the Middle East, which was still at war.

When I was posted as a major to the Sëcuritë Brigade on the aircraft carrier *'Clemenceau'*, *I* was admitted to the naval officer corps through a competitive examination, in the branch specialising in sëcuritë and the environment. In order to work on aircraft carriers and naval bases as a safety assistant, I did a final refresher course in nuclear prevention at the Ecole Atomique in Cherbourg and at Cazaux in the French air force. After the Falklands War and the conflicts in the Persian Gulf, civil defence and security doctrines had changed, and I wanted to gain experience of working with the public. When I was posted to the Marseille Marine Fire Brigade, I was to be confronted with the everyday risks faced by our fellow citizens, with industrial and road traffic accidents that were very deadly, and with terrorist threats, at a time when the war in Iraq was once again raising the spectre of chemical weapons, bomb scares and attacks that were multiplying...

My nuclear physics teachers, the naval doctors and the highly respected radiation protection major 'JJD', had passed on to me a passion for studying nuclear sciences and biology applied to ionising radiation. It was the time of the major technological disasters of Bhopal, Chernobyl and Torrey-Canyon that made me aware of environmental issues. Under the influence of Jacques-Yves Cousteau and Rene Dumont, a media-driven and politicised

---

1 My faithful assistants, sailors Jean-Pierre Sylvestre and Sylvain Mahuzier, became brilliant naturalists, authors and confërenciers, tireless guide-explorers of Antarctica, Cape Horn and Tierra del Fuego.

ecology was taking off, and I was also paying close attention to the application of the treaties on the prohibition of chemical weapons and the limitation of nuclear weapons, which were subject to international controls in an attempt to control their threatening proliferation...

I had соттепсё a ёtudier la Шёопе de Involution a bord d'un porte-avions, apres une rude jouriree, de postes aviations, de poste de combat et d'exercice de sёcuritё, je reglais les affaires courantes liёes a maitre-adjoint aupres du capitaine de fregate, chef du service sёcuritё. Between two night shifts, finally isoM in my bedroom-office, I had a little hour to prepare for my naval officer exam and read a few pages before falling asleep. The works of Konrad Lorentz, Henri Laborit, Ernest Mayr, Georges Gaylor Simpson, Albert Einstein and Richard Feynman, which I carried in my little sailor's suitcase, helped me to get through these long voyages, in a kind of intellectual escape that was to become increasingly important.

6. Training of ёlёves-officiers spёcialisёs on board a goёlette

In 1983, I met Professor Patrick Tort during a conference on Darwinism and society in the late nineteenth century. When he founded the Charles Darwin International Institute in Puycelsi (Tarn), I took part in the writing of the long preface to Charles Darwin's 'Logbook' for the maritime and exploration section. My experience as a sailor and my research into the history of maritime exploration proved useful in trying to uncover the beginnings of his lost theory on board the H.M.S. 'Beagle', in the company of the officers and sailors on the hydrographic mission that led to his discovery.

At the end of this highly fruitful collaboration, we drew up a project inspired by the exploration methods of Charles Darwin and the crew of the H.M.S. "Beagle". We drew up the initial specifications for a prototype of a multi-mission, long-duration sailing vessel for scientific exploration in the field of evolution and civilisation, designed to carry out uninterrupted survey missions simultaneously at sea and on land. We have submitted it to the Academie de Marine and communicated it to the Ministёre de l'Enseignement Superieur, de la Recherche et de l'Innovation, and we have sent this bottle to the sea to the Tara association and to several institutions likely to implement this maritime and scientific heritage (IFREMER, IRD...).

7. Prototype of a mixed scientific exploration sailboat.

I had travelled the world and made a few trips that gave me the opportunity to write a four-volume naturalist suite and three essays on the complex processes involved in biological evolution and the transformations of civilisation, including this one, which was a must on low biological energies, very complementary to my research and numerous questions on the cutaneous respiration of Lissamphibians, observed in the Pyrenees and Cevennes.

8. Red frog and Euproctus in the Hautes-Pyrenees.

The power developed by States to maintain their socio-economic activities and guarantee the defence of their interests has become hypertrophied in order to ensure the continuity of supplies of energy resources and maintain the integrity of territories against any aggression. As a marine historian and naturalist, I was able to measure the gap between the low biological energies at work in an organism and those deployed by our extreme productivist societies, in search of limitless growth, an inevitable source of decline.

I had discovered the complexity of these issues on my travels and during my naturalist explorations. To begin my fundamental and applied research, I had to develop a method for analysing the complex processes involved in biological evolution and the transformations of civilisation.

Initially, I set out three guiding principles:

1-   The divergent transition, between linear and non-linear.

2-   Adaptability as a stochastic revolution operator.

3-   Irreversibility: complex evolutionary processes.

It is a truism that the phylogënëtic trees that summarise the evolution of species since their origins are split into a multitude of divergent trajectories. Charles Darwin noted the importance of this in his pencil sketch of his theory of species (1842) and in The Origin of Species by Natural Selection (1859) he traced very explicit ëvolutive sclrenmis to his understanding of the complex^, which he expressed in this sentence.

"In the course of a thousand generations, differences of infinitesimal
infinitesimal smallness must inevitably count".

What will later be called the sensitivity to initial conditions that contributes to the divergence of evolutionary trajectories during the adaptativeness of species since their origin, in his last paragraph, he concludes as follows:

"If we think that these forms [Speaking of species] so admirably constructed
so admirably constructed, so differently conformed, and so
dependent on one another in such a complex way [...]
Law of the multiplication of species [...] which has as its consequence
natural selection, which determines the divergence of characteristics [...] The extinction of
forms [...
characteristics [...] The extinction of forms [...]
from such a simple beginning, never ceased to grow and grow
and is still developing".

In paleontology, Georges Gaylor Simpson, who contributed to the synthetical theory of revolution, effectively included the variations of a species, while schematising the constraints of natural selection, in his representations of the revolution of horses, sabre-toothed tigers and ammonites. By making successive cuts at the points of divergence, he demonstrated the various modes and rhythms of biological revolution.

At another level of resolution, the physicist Richard Feynman devised divergent diagrams to resolve the interactions of atomic particles, both wave and corpuscle, a process that he skilfully assimilated into quantum field theory.

In addition, the statistical physics of Lev Landau and Evgueni Lifchitz gave me access to the domain of the visible and of sounds perceptible by the senses. In this field of mesoscopic transition, I was interested in metastable bodies, looking like nematic and cholesteric pseudo-crystals, comparable respectively to biological membranes and protein complexes, the essential constituents of the cells that make up an animal.

Two major concepts would computerise my exploratory mëthod of complex processes in biological ëvolution, necessarily ëextended to the transformations of civilisation.

- The first of these took into account the theory of units of level of integration by the Spanish biochemist Faustino Cordon (1909-1999), which gives a primordial role to the basic proteins at a directly supra-molecular level from which the cell (supra-proteinic) and the animal (supra-cellular) emerge. His close and late collaborator Chomin Chunchillos and his daughter Teresa had come up with some very interesting clues as to how the cell pulsates and how cellular activity is coordinated, which I found intriguing. In dealing with consciousness, Faustino Cordon had suggested the notion of strata, which I evoke in the fourth volume of the naturalist suite *"Indri indri, voyage aux origines de I'Humanite"* and in the essay *"La septieme corde", the* ultimate synthesis of complex evolutionary processes. It turns out that these strata are very significant and observable during divergence processes, particularly during the transition between linear and non-linear. Here I found a powerful field of exploration that called on several scientific disciplines and required me to make serious efforts to understand them.

- The second concept, and by no means the least, consisted of using adaptive trajectories to explore these adaptive strata in order to reach G.G. Simpson's integrated model, and on an experimental basis: daring to introduce the reversal effect (Tort, 1983), specific to the human species, whose social instincts had to be traced back to their animal origins, without delving into the mysteries of sociobiology.

9. Ginkgo biloba leaves,
natural vein divergence

My starting point for this essay was Henri Laborit's publication *"Les comportements"*, in which he highlighted the vital role of the joint circulation of protons and ёкйго^, dangerously alkrable during a state of stress or shock. At the interior of a biological cell, in the inner membrane of the mitochondria, prokine complexes jointly ensure the transport of ёк^го^ and the translocation of protons, contributing to cellular respiration.

Before tackling these uniks of biological integration level, we describe these not-so ёlëmentary particles that actively participate in the relative vital cohesion of organisms, and in the stochastic adaptativik of species during their contingent evolution.

The textbook approach applied in the field during my explorations led me, when observing each species, to take account of the adaptability relationship between the fundamentally random variations - genetic and epigenetic, morphological, behavioural and, by extension, social - in their relationship to fluctuations, to chance, and to natural and anthropic selective constraints, which are highly correlated. The stochasticity that results from this relationship of adaptability, a source of unpredictable divergences, was to become the subject of my research into the complex processes of biological revolution and the most dramatic transformations of civilisation, which I had experienced intensely.

As we suggest, in the urgency of the civilisational and ecological crises, by constantly studying the critical points of these bifurcations, globally and locally, shouldn't we be moving closer to the ways in which low biological energies function?

To do this, we need to open a few doors using the keys of quantum physics, by diving deep into the integration units of living organisms, from the level of proteins to that of the cell and the animal. Various species have developed social instincts that, in the case of human beings, have given rise to imaginative states of consciousness that require us, in the urgency of crises, to go to the limits of our intelligence to safeguard the fragile edifice of civilisation: to preserve humanity, to safeguard species and their environments.

Chapter 1

Protons and electrons, the keys to life.

Quantum physics describes the laws of radiation and those of the particles that make up the atoms assembled into molecules in inert matter and in living organisms that differ from them. These are the proton (Ernest Rutherford, 1919), the neutron (Ernest Rutherford, 1920) and the electron, to which we must add the concept of the photon (Lewis, 1926). Through its action, sunlight triggers photon synthesis in plants by activating the dissociation and circulation of electrons and protons in the cell. These two fermions are the keys to life.

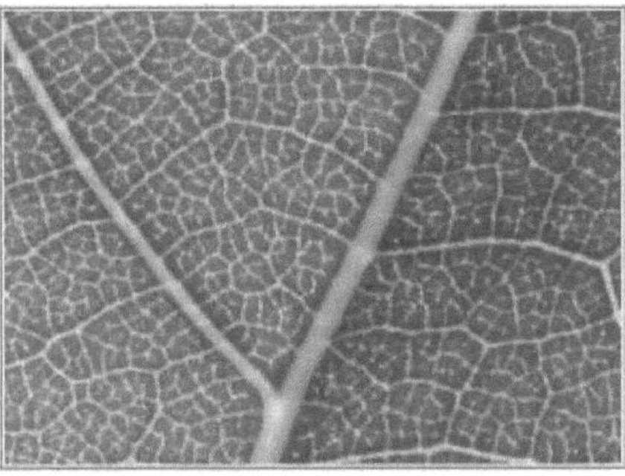

10. Plant cells penetrated by the sun's rays

In animals, this level of resolution of cellular activity takes place in the cytoplasm, which contains organelles called mitochondria, within which ёкйго^ and protons are translocated via complexes of т3ёгёз protëins in its inner membrane. In this organelle a transition takes place that can be explained with the help of quantum field physics and statistical physics, moving from the realm of the indiscernible to the discernible, to that of the observable by a naturalist, within the framework of the mëthode of analysis of complex ëvolutive processes. Before describing cellular respiration, we present the proton and the lklectron, which are very involved in the first-level unit of biological integration, the prokine.

11. In Bhutan, the Takin (**Budorcas** taxicolor) feeds on vëgëtals.

Albert Einstein (1879-1955) ёпопсё the thc'ories of restricted relativity[2] and дёпёгак[3]everyone knows his сёкЬге formula which relates energy to the mass of matter by the square of the speed of light:

$$E = M C^2 \qquad\qquad [\,1.1\,]$$

In 1905 he риЬНё papers on Brownian motion and restricted гекй^ё by ёmaking hypotheses about light quanta.

"The surgeon and botanist Robert Brown (1773-1858)
took part in the exploration of the Australian coast aboard the H.M.S
Investigator commanded by Matthew Flinders, from 1801 to 1803.
Using the microscope, the botanist discovered the nucleus of
the plant
cell and described the random movement of
pollen grains that bears his name. Robert Brown[4] was a
correspondent of Charles Darwin, the young naturalist visited him
to try out his microscope. Clearly he was not keen to
reveal his discoveries to Charles Darwin, who confided in
him in his autobiography,
Brown's distrust of him".

Albert Einstein has rellё the duality of light, simultaneously undulatory at the nanoscopic scale, which manifests itself from the microscopic to the macroscopic in a quantified corpuscular form in organised structures in inert or living matter. They appear more coherent and are perceptible to us, through our senses, nervous system and brain, mediators of our sensations perceived and interpreted by our consciousness, according to our knowledge and experience.

---

2 Restricted relativity boils down to two postulates according to which the laws of physics have the same form in all Galilean rёfёrentials. The speed of light in a vacuum has the same value in these rёfёrentials.
3 The relative gёnёrale dёcrit the motion of a mass subject to gravitation from which results an inertial motion in a space-time тотЬё by these masses.
4 After his trip, Robert Brown published a work on the flora of Tasmania and Australia (1810), for which Mount Brown is named.

12. Sun photons, wave and corpuscle

Wolfgang Ernest Pauli (1900-1958) established the exclusion principle according to which a particle cannot occupy the same quantum state, he discovered the spin of the nucleus and explained the hyperfine structure of atomic spectra. As for Werner Heisenberg (1901-1976), his uncertainty principle tells us that quantum phenomena are fundamentally random, statistical and probabilistic in nature. These two scientists contributed to quantum field theory (1929).

Whole-spin bosons (photons, gluons, Z, W, Higgs bosons, and mesons...) that satisfy the Bose-Einstein statistic do not satisfy the exclusion principle. They can occupy the same quantum state at low temperature, a state conducive to superfluidity and superconductivity.

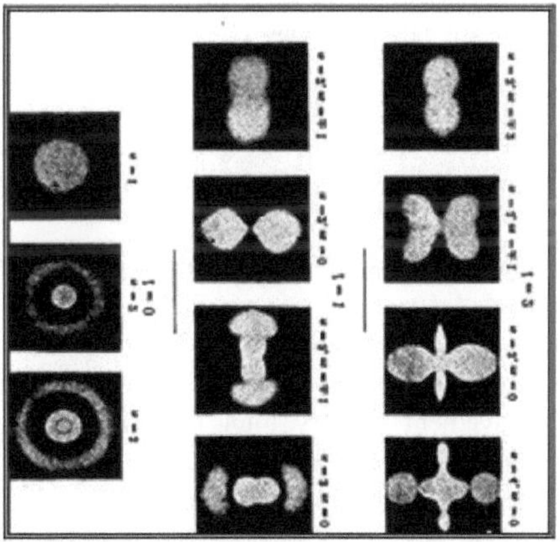

13. Representation of a hydrogen atom
at different energy levels (H.E.White, 1931)

A W. E. Pauli should be associated with another Darwin, Charles Galton (1887-1962), grandson of Charles Darwin and son of George Darwin. This little-known physicist calculated the fine structure of the hydrogen atom, and published *"The Wave Theory of Matter"* (1930) and *"New Conceptions of Matter"* (1931). This former soldier, who was heavily involved in the last two world wars, was also interested in the problems of world population, and wrote *"Le prochain million d'annees"* (1952). In this book, he discusses overpopulation and eugenics, undoubtedly influenced by the work of his uncle Francis Galton and the heavy legacy of Charles Darwin, whose theory of the origin of species through natural selection had suffered the clumsy excesses of Social Darwinism.

In 1927, C.G. Darwin introduced the corrective term of the Pauli-Darwin equation, which establishes the random interaction between the atomic nucleus and the electron. Here we are at the heart of the problem of oscillating perturbations caused by fluctuations in nuclear interactions.

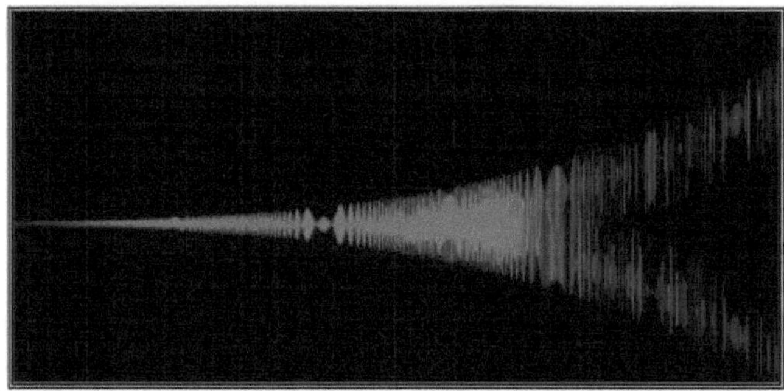

14. Simulation of a divergent wave "particle".

The motion of a free particle (proton, electron, ion) as described by Paul Dirac (1928) is the superposition of two motions, which we simulate above.
- Uniform translational motion.
- An oscillating movement.

To these must be added, in the space-time of its resolution, an eventuality of divergence, formalised in the diagrams of Richard Feynman (1918-1988) to represent the interaction of particles.

The first two terms of the Dirac equation characterise translation at constant speed, and it is the third oscillating term that gives rise to the *"Zitterbewegung"* (Schrodinger, 1930), whose oscillation amplitude is[5] is inversely proportional to mass, like the Hawking temperature for black holes.

These states of negative energy are seen in the Hawking effect, when tiny amounts of radiation are emitted by black holes.

The origin of this oscillator is thought to be an interference between the positive and negative energy solutions, such as those we have just described between the proton and the electron.

This discrete oscillatory movement at very high frequency is manifested by divergent amplitudes with a chaotic tendency, apparently regular at a lower resolution, a phenomenon observed in calcium ions (Miyazaki and William N. Ross, 2017).

The wave function is dëcrite by the no less cëlebre ëquation lieaire of Erwin Schrodinger with which we considerëre that an atomic particle placëe in a potential reacts like a wave

---

5 Depending on the angular frequency of the oscillatory component, the energy of the quantum harmonic oscillator is E = Йю/2, й is Planck's reduced constant, и is the wave pulsation which depends on its frequency.

during its ëvolution in time. If the potential comes from an electric or magnetic field, or from light as we have just mentioned for pliotosvntliese, this interaction makes it possible to "*see*" the consequences of this perturbation.

$$ih\, \partial\psi\, /\partial t\, (x,\, t) =$$

$$-(h^2\, /2m\,{}^*\partial^2\psi)\, /\, ((\partial^2 2x\, (x,\, t) + V\, (x)\psi(x,t))\, :\, \forall x,\, t \qquad [\,1.2\,]$$

By applying this non-relativistic ëquation, we obtain the instantial modification of the wave which depends on the mass of the particle and the potential forces $V\,(x)$ experienced by it. To estimate the energy of the interactions we have an operator, the Hamiltonian, which corresponds to the total energy of the svstëme, for a particle of mass m and momentum p, constrained by a potential $V\,(x)$:

$$\hat{H}(x,\, p) = (p^2/\, 2m) + V\, (x),\ op\acute{e}rateur\ Hamiltonien\ de\ l'\acute{e}nergie.$$

$$p = -ih\,{}^*\partial\, /\partial x,\ op\acute{e}rateur\ d'impulsion.$$

$$x,\ op\acute{e}rateur\ position. \qquad [\,1.3\,]$$

$$H(x,\, p) = (p^2\, /2m) + V\, (x),\ \textbf{Hamiltonian energy operator.}\ p = -ih\,{}^*d\, /dx,\ \textbf{impulse operator.}$$

x, *position operator.* [1.3]

Schrodinger's non liiwaire liquidation is less well known, we will have the opportunity to evoke it in the form of Eugëne Gross and Lev Pitaevskii's liquidation in our approach applied to Bose-Einstein condensates predicted in biological ënergies at low tempërature. More specifically, to glimpse the possible ëquences of a saturation proton field, pu^ a threshold, on the conductivity, fluidity and permëabilitë of membranes associated with protëine complexes.

$$i\partial t\Psi = \Delta\Psi + \Psi(1 - |\Psi|^2) \qquad [\,1.4\,]$$

Before beginning the reconciliation between the first two units of integration levels of the living, from proteins to the cell, I would like to comment on the main states between the nanoscopic components, of the non-living atom, whose units differentiate and emerge in vital coherence, during the mesoscopic transition, from the microscopic to the macroscopic. To this end, we are using quantum mechanics linked to statistical physics applied to the biochemistry of living organisms and to the evolutionary biology of species, as a complex process, as proposed in our definition of adaptability.

The electron, the proton and the photon are very much involved at the level of the basic proteins, they contribute to the dynamics of the cell's activity, and consequently to the vital coherence of the units of integration level, proteins, cells, animals, and to the revolution of the species whose variations are confronted with the selective constraints of their environments. We are used to considering these levels as hierarchical, which is justified when we observe the organisation of living organisms, each emerging level being different from the previous one, while remaining dependent on the previous one, in a complex network of interweaving. A mosaic of interactions usually described from the genotype to the phenotype that differentiates each species whose population evolves through its '*adaptability*' in space and time.

However, at the level of the particles described by quantum physics, Hermann Minkowski's (1864-1909) notions of space and time merge into a fourth dimension that escapes our perceptions.

While most of the biological and ecological processes that contribute to the revolution of species take place in the finite spaces of organisms, species and their ecosystems, in a one-way time...

These two paradoxical notions are fundamental to the rest of our presentation, insofar as particles, as notorious waves, are linked together by the energy flows of their quantum fields. Their couplings and interferences manifest themselves as a function of their energies and are translated into a quantum arrangement of corpuscles that make up the atoms that make up our reality. These energy levels can be identified by a spectrum of fine structures, specific to each chemical element classified in the Mendeleiv table. It is thanks to these quantified interactions that biochemical combinatorics contributes to the unity of levels of integration of living organisms temporarily structured by the establishment of a vital coherence. The lives of the individuals that make up a species are effectively dependent on time for a limited period specific to each species. This obliges us to examine simultaneously and in complementarity the atomic level, if not the subatomic level, dynamic but not living, to ensure the delicate transition between the units of the highly integrated and successively emerging biological levels, from proteins to the cell, right up to the animal.

In quantum physics, Richard Feynman's divergent diagrams, in space and time coordinates, show this property explicitly. I often quote the creation of a pair of a proton $(p+)$ and an anti-proton $(p-)$, produced when radiation of sufficient energy interacts with the field of an atom. On the ordinate, we find space, where the more or less inclined trajectory of the wave and particles is plotted as a function of their speeds, and on the abscissa, time, as the waves and particles move forwards and backwards in this space, relative to the direction of time. These divergent diagrams are associated with very rigorous imitlieimite demonstrations that belong to quantum field theory, where space and time are indistinguishable, and which are outside the scope of this naturalistic essay.

Our problem is as follows: beyond the apparent hierarchy of integration-level units (protein, cell, animal), is there in fact an underlying unit that expresses itself in space-time in the form of a fundamentally random discrete wave?

Could this unit be involved in the coordination of cellular activity by the presumed action of a pulsed quantum field of highly excited protons, which at a threshold would act punctually on the rearrangement of the atoms of the cell and its constituents, in a confined space with a relativistic tendency, with biological effects that are ultimately dependent on time?

Having said that, it is not a question of systematically operating a reduction to the atomic level alone, but of sensing the transition between different areas of understanding of non-living and living matter that obey concepts long attributed, through ignorance, to mystified vital forces, such as a divine creative power. It's easy to see why our approach needs to be approached with caution, as the words and theories of quantum physics are used to conceal pseudo-scientific methods and unsavoury therapeutic practices that tend to pit soft medicine against hard medicine.

The ancient practices of magnetisers and healers use skilful manipulation to correct physical and psychological deficiencies. These empirical practices are based on a different way of thinking. The Tibetan Amchis and Chinese doctors have developed ritual procedures in a typically Eastern system of thought that uses low energy, the vital energy known as *'Chi'* in Asia and the Sanskrit name *'Prana'* in India.

These male doctors are able to bring relief to their patients by taking into account a range of individual, family and social factors relating to a disease, which they skilfully integrate into their consultations. After taking a fairly complex pulse at several points of contact, where they apply judiciously modulated pressure in direct relation to the organs affected, they make an initial diagnosis.

This is followed by a careful examination of the patient (tongue, ears, eyes and skin) to locate the critical points where energy is concentrated in superficial circuits, called meridians, linked to the internal organs affected by the disease.

To rebalance these energies, known as *'Yang and Ying',* a kind of duality of positive and negative energies, the aim is to dissipate the surplus or restore the lack of them through localised movements, during appropriate massages or with the help of acupuncture practised in various forms, needles, heat, fumigation. These mystified shamanic practices are often combined with the use of medicinal plants or mineral and animal mixtures, stimulated by a spiritual atmosphere that invites the psyche to participate in the healing process. The patient is invited to recite mantras, repetitive sacred prayers adapted to each case, and to practise ritualisës hand movements.

These gestural mudras and the low-toned sacred chants are performed in soothing meditative postures that are said to modify states of consciousness to the point of promoting healing, or at least ensuring well-being.

These traditional medicines and rituals have been formalised for centuries and correspond to each pathology, treating both the physical and psychological aspects of the patient.

They are mentioned in sacred writings from India and Tibet. In particular, those taken from the Upanishads, a long narrative written more than five hundred years before our era. I quote the end of Karika IV, the extinction of the dying firebrand:

> *"Having attained knowledge to the best of our ability,*
> *we salute the obscure, the profound and the integral,*
> *eternal and pure identity,*
> *Which is the abode of Unity".*

We are at the birth of gods and goddesses, in a quest for the indiscernible, mystified by the beliefs of the Hindu and Buddhist schools, and of ancient Greece with its different atomistic conceptions. The sacred texts of religions with well-established dogmas used the miracle attributed to their divinity, summoning a mediator God, whose religious representatives firmly opposed any discoveries that dared to defy them by disturbing their authority and the social order they advocated, at the whim of the princes and kings who abused them. So much so that scientific practices that did not conform to these truths sent many witches and scientists to the stake of the Inquisition, and still justify murderous acts.

Indian ayurveda and the practice of yoga are ways of life that attempt to control body energy in a preventive way through diet and physical exercise directed towards each type of organism according to a codification that escapes western medicine. Long opposed to each other, these two forms of medicine are now tending to computerise, and research into ethnology and quantum physics applied to biology are making it easier to identify the deceptions of charlatans and to try and understand these *'magical'* healing powers.

Hindu asceëtes can withstand the heat, and those in Ceylon who walk on fire recite mantras to cool them down! Tibetan monks practise *'Tummo'*, exposing themselves to the cold with total denudation, according to Alexandra David-Neel's testimony. Through a few meditative gestures and postures, which have nothing mysterious about them, as I was able to observe, they manage to produce physical warmth, integrated at a psychic level in a surge of spirituality that mobilises states of consciousness.

15. Over 3,000 metres, internal heat.

These texts quote:

> *"The subtle energy accompanied by a pleasant warmth*
> *that begins to penetrate every atom of the body.*

In the last book of the Path of Wisdom and the Yoga of Emptiness, reference is made to the existence and non-existence of atoms, which are nothing more than a reflection of our own minds...

After this parenthesis, let's return to the firm continent of Science, and of quantum physics, which is only as subtle as the harsh ëquations and rigorous expëriences of the physicists and biologists who are conducting promising multidisciplinary research around particle accelerators, judging by the many theses I have had to explore to argue my proposals.

If we consider that the interaction of the proton and the electron are the keys to life through the intermediary of a ***structuring and/or destructuring*** nuclear binding energy, which would manifest itself in the cell at a quantum level, discrete and random, by definition. This free ***nuclear F*** energy causes disturbances in the integration level units (proteins, cells, animals), and their fluctuations are associated with thermal and hydric effects in a dynamic that causes dissipation of this energy in the biological environment. It is during these complex, highly integrated processes that the specific arrangements of each unit - proteins, cells and animals - emerge, periodically subject to structural and functional rearrangements, under genetic control and during complex epigenetic interactions, whose morphogenetic variability is effectively under the constraint of their sëlective environment.

A priori, according to the hypotheses suggestedërë later in our exposë, cellular activity would be coordinated by a unit or rather a field of protons whose action would be faster than those of the biochemical interactions that take place in the units of biological integration levels.

This hypothesis raises important questions about the states of vital coherence during the contingent revolution of a species, whose stochastic adaptability depends on its fundamentally random variations in their complex relationships with the selective natural and anthropic constraints that arise at random from circumstances, influenced by our socio-economic activities!

This proposal requires us to describe the proton and electron, and their binding energy, as an observable of their interactions. With the help of quantum field theory, we move away from the usual representation of the particle, long assimilated to a matërial point. In what is known as space-time, concentrations of energy are formed to which physicists give wave properties that, according to Werner Heisenberg's uncertainty principle, can be dëcribed in the form of a particle. The restricted meaning of corpuscle, which has contributed to a reduction to the atomic level alone, no longer applies to the notion of quantum field, which encompasses an ensemble of production and disappearance, as well as the quantified distribution of energy levels into fine, and incidentally hyper-fine, structures. These indiscernible strata are the discrete random variables of discernible chemical adjustments, and of observable structuring and/or destructuring biochemical affinities, and consequently, one of the causes of the essentially random biological variations that are expressed at random, under the effect of selective constraints during the contingent evolution of a species, by :

Stochastic adaptability.

Each atom is characterised by a spectrum, a sort of signature that represents the distribution of its energy levels into fine structures sensitive to the action of an external field. When an atom is perturbed, depending on its susceptibility to a magnetic or electromagnetic field Bo , transitions appear to the experimenter between the fine and hyperfine structures of the spectrum of the atom in question.

It is this quantum field that is the common thread running through our research into low biological energies, but first we need to answer a key question: how do atomic nuclei and their electrons interact?

To simplify our presentation, we will use the most abundant atom in the Universe, with a natural abundance of 99.98%: the hydrogen atom, made up of a positive proton ($H+$) and a negative electron (*e*).

The proton, noted (*p+*) by physicists, is a fermion with spin 1/2, its mass is 1.67 262 192 369 $*10^{-27}$ kg, according to [1.1], expressed in energy, i.e. 938, 2 720 813 Mev/c^2 . Its lifetime is greater than $2.1*10^{29}$ years, and its positive electric charge, which is not homogeneously distributed, is 1.6 021 766 208$*10^{-19}$ Coulombs for a radius of 0.84 184 fentometres.

According to quantum chromodynamics, the proton is made up of two up quarks (*u, u*) and one down quark (*d), and* its binding energy corresponds to the kinetic energy of the quarks combined with that of the moving gluons. The pressure at the centre of the proton is at its maximum, 10^{35} Pa, exerting a repulsive pressure from 0.6 to 0.8 fentometres, then low above 2 fentometres, as the inverse confinement pressure.

When the quarks move apart, this freedom increases their interaction, producing a quark-antiquark pair. The proton's three quarks (*u, u, d)* are bathed in this *"turbulent sea"* of quark-antiquarks.

Stemming from the fleeting materialisation of a gluon, strange quarks (*s)* also affect the proton's magnetic moment. A priori, the charge of the proton *(p+)* does not change in this temperature from which the strange quarks (**s**) arise, but it is distributed differently in its volume, modifying its charge radius and its magnetic moment. While its instability is extremely low, in its form as an excited *H* ion$^+$ in search of electrons, the proton is very reactive to its own field configured in several energy levels likely to modify the arrangement of neighbouring atoms according to their susceptibility, which in turn would piëger the proton field by transforming its condensed constitution to a certain extent.

Thus trapped in a restricted space, it would manifest itself at low temperature in a double state, of protons and virtual photons, conducive to this other condensed state, close to a pseudoplasma. Could this state occur in the cell, in the intra-membrane space of the mitochondrion?

You will understand that, as a naturalist, this question is more about reflection than assertion...

The spectrum of the hydrogen atom can be broken down into several energy levels, starting with the fundamental level.

- The Theodore Lyman series in the ultraviolet wavelength range[6].
- Johann Jakob Balmer's series is in the visible domain[7].
- The sërie of Ritz-Friedrich Paschen[8] corresponds to the infrared.
- Then come the borderline sëries of Frederick Brackett and August Pfund[9]
- Then, those of Curtis Humphrey and those discovered by Peter Hansen and John Strong, when the clectron moves away from the proton.

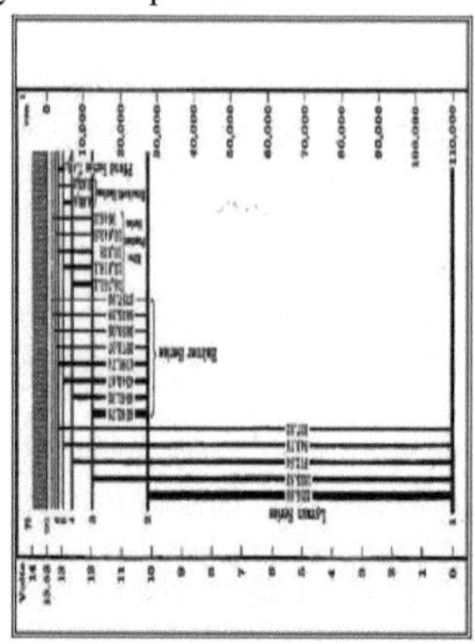

16. Energy levels of the hydrogen atom
after W. Gotrian[10], 1928.

In 1947, experimenting with radio frequency spectroscopy to cancel out sensitive Doppler effects on 1 hydrogen atom, Willis Eugëne Lamb (1913-2008) and Robert. C. Retherford showed that the 2 level^2 sj/2 has an energy 1,000 Mhz greater than the $_{22}$ level p1/2 .

The physicist Hans Albrecht Bethe[11] (1906-2005) commented on this displacement of spectral lines or Lamb shift by the interaction of the hydrogen atom with the quantum radiation field, whose non-zero energy of virtual fluctuations *would "diverge"* by disturbing the real particles. The evaluation of this *"Lamb-shift"* was relativistic and required a resolution using the

---

6 The Lyman series is understood from its fundamental state n=1/0 A/Ovolts, for n=2/ 1215 A/10.2 Volts, for n=3/1025.83 A/12 Volts, for n=4/972.54 A/12.6 Volts, for n=5/949.76 A/12.8 Volts, for n=6/937.82 A/13.6 Volts.

7 The Balmer series starts at level n=2 for n=3/6562.79 A/12 volts, for n=4/4861.33 A/12.6 volts, for n=5/4340.47 A/13 volts, for n=6/3970.07 A/13.6 volts, for n=7/3889.05 A/13.70 volts, for n=8/3835.39 A/13.45 volts, for n=9/3797.90 A/13.49 volts.

8 The Ritz-Paschen series starts at n=3, for n=4/18.751 A/12.6 volts, for n=5/12.818 A/ 12.8 volts, for n=6/ 10.938 A/13.2 volts, for n=7/ 10.049 A/13.4 volts, for n=8/ 9546.2A /13.45 volts.

9 The Brackett-Pfund series is at n=4, for n=5/4.05 pA/13 volts, for n=5/2.63 pA/13.2volts. from n=5 for n=6/7.4 pA/13.2volts.

10 Graphische Darstellung der Spektren von atomen und ionen mit ein, zwei und drei valenzelek-tronen, W. Gotrian, Springer, 1928.

11 His research into the rejection of electrons by metastable atoms and quadrupole interactions in molecules, which we will discuss later, should be noted.

formalism of quantum electrodynamics to reduce the errors and increase the precision rendered a :

1 057.61 +/- 0.16 Mhz (Appelquist and Brodsky, 1970).

In chapter three, I based myself on the experiments of the physicist Michele Glass-Maujean (1974) relating to the method of anti-crossings, the n=3 and n=4 levels of the excited hydrogen atom formed by dissociation of the molecule[12] *H2*, a component of "*still water*" that is very present in the cell in its dissociated form as electrons (*e*) and protons (*н+* or *н₃о+* in aqueous solution), which are involved in cellular respiration in the mitochondria.

Not only has the researcher improved the precision of these frequencies, but she has also located the values of the quantized radiation field that would be divergent, and has determined the different speeds of excited *н+*, to very excited *н+**. Other scientists have described very precisely the splitting of the hydrogen atom's energy levels into hyperfine structures, at the limit, rendered in this ëtat of very excik *н+** ion, whose ëtats in a соиПиё medium at low tempёrature, were questioning us. In this condensable ëtat, would this quantum field of protons be comparable to a micro-plasma of virtual photons and protons, piёgёs in a nanoscopic closed space?

The ëlectromagnëtic field стёё by these very exc^s *н+** protons produces a dёcalage of the energy levels of the ёк^го^ of the neighbouring atoms which are more or less blindёs, therefore likely to react differently to the variation in the proton field. When it goes from weak *н+* to excessively strong *н+**, it would cause, depending on the size of their ëlectronic layers, dёplacements of their energy levels, if not, ёjections of electrons up to ionisation, or the creation of ëlectronic pairs, as well as the formation of free radicals...

The fine structure constant *a* is the dimensionless ratio of the mass of the proton *mp* to that of Election *me*, the mass of the proton depends on the laws of quantum chromodynamics ёvoquedёcёdently and that of the lklectron of the Robert Brout- Frangois Englert-Peter Higgs (*BEH)* field associated with the Higgs boson, whose field[13] confers mass on the bosons and fermions; we must also consider the Yukawa couplings[14].

The fine structure constant *a* dё also depends on the inhomogeneous ёelectric charge of the proton, a discrete nuckar variable that dёetermines the intensik and akatory repartition of the ёlectromagnëtic force. The mass of the proton is therefore related to Гёдшуяк^ of the fine structure, by the strong nuckar forces of the ёchangёs gluon fields between the quarks: We must considerё the effect of variations in the constants on the energy levels occasioned by the frequencies of the ё photons emitted or absorbed during more or less close interactions with the electrons of the atoms or between the excited protons. The principle of non-localisation '*Quantum entanglement*' (Shrodinger, 1935) experimented in the 1980s by the French physicist Alain Aspect takes on its full importance in our demonstration of the coordination of cellular activity by a presumed quantum field of threshold pulsed protons. Its celerity would become faster and faster, reaching relativistic speeds, than the reactions that determine the biochemical and thermohydrodynamic events of the biological cell.

The fine structure of hydrogen comes from the kinetic moment of the electron and proton,

---

12 The hydrogen molecule н2, measures 2.3 10-8 cm, its binding energy is 4.72 ev, energy must be supplied to dissociate Щ, whereas the association of two hydrogen atoms is spontaneous and releases energy. At 27°C its energy is 0.038 ev, and at 0°C its mean square velocity is 1.85 ms^{-1} .
13 At low tempёrature space would fill with Higgs particles, the bosons thus acquiring an effective mass.
14 The Japanese physicist Hideki Yukawa has shown the mёdiator role of a pion in a Coulomb potential during the coupling of a mёsonic field and a fermionic field, as well as the resulting attractive forces, i.e. 10^{-15} m for the proton and a mass of 140 Mev for the mёson.

which can be modified by its charge linked to the internal constitution of the turbulent sea of quarks, their angular momentum or spin being :

$$i = +1/2$$ in two states.

- Without a magnetic field, the magnetic states are at the same energy level, degenerate.
- Constrained by a B_o magnetic field, by the Zeeman effect, the *2+1* random energy states undergo a spacing proportional to the gyromagnetic ratio and the strength of the external magnetic field and take on two orientations:

$$m = +1/2$$

$$m = -1/2$$

Atoms resonate at different frequencies with deviations of several megaherzts, for example under the influence of a flow of electrons or a field of excited protons, in the case of the mitochondrial environment and the cell. The fine and hyperfine structures of atoms, and hydrogen in particular, are due to the electron's kinetic moment, *5* denoting the electron's spin magnetic number *(+/- 1/2)*, for *n* the principal quantum number, *l* the azimuthal quantum number and *m* the magnetic quantum number.

The proton has the same property for its two magnetic states *m = +1/2* and *m = -1/2*. The nuclear spin depends on the number of protons and neutrons in the atom, the intrinsic angular momentum *I* generates a magnetic field, and a magnetic moment *u* linked to the nuclear spin moment by the gyromagnetic ratio *y* , intrinsic to the nucleus. When the proton in the nucleus of a hydrogen atom is subjected to a magnetic field of 2.3 488 Tesla, the observed frequency is 100 Mhz and its *y* is 26.7 519 rad T s^{-1-1} .

Let's look at the interaction of the proton's magnetic field with an external magnetic field *Bo.* Without an external magnetic field, the states of the proton are at the same degenerate energy. By the Zeeman effect, the external field B_o interacts on the *(2i+1)* random states, *i.e. (i = +/- 1/2)* the energy *E = - my h $_{Bo}$* is spaced from *AE = $_{hBo}$* which takes two orientations.

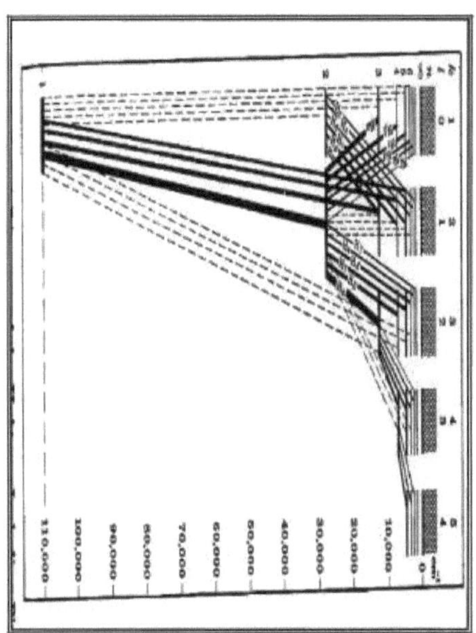

17. Hyperfine structures of hydrogen Γatoms,
crossings and transition levels, after Gotrian, 1928

This effect, identified by Pieter Zeeman (1896), produces subdivisions of the atoms' energy levels as a function of the Be magnetic field. If the number of ultrafine lines in the spectrum obtained is odd, the effect is described as normal; if it is even, it is considered abnormal. Depending on its strength, the field simply causes a ë thickening of the lines in the initial atomic spectrum.

While the fine structure of hydrogen lines is relatively homogeneous, due to the elliptical trajectory as a function of **k**, the electron's azimuthal quantum number, the fine structures are modified and can give rise to forbidden transition zones, groupings of energy levels and extremely complex line crossings. This led us to consider these random repercussions at the level of ultrafine structures caused by a magnetic field that would produce this discrete splitting of energies.

In particular, when the electron moves away from the proton, in the excited, isolated $H+$ ion state, or forms a field in an acidic medium that goes from weak to strong $H+*$, producing the presupposed underlying divergences in the quantised field. We need to consider the consequences of the distribution of different proton velocities over the range of frequencies that result from these strong interactions at saturation. During *s-wave* proton-proton scattering at low energies, J.P. Naisse[15] underlines.

> ***"To the electromagnetic amplitude cannot be added***
> ***- simply - an amplitude***
> ***induced by strong interactions"***.

An additional term in quantum electrodynamics would give rise to discrepancies, with

---

15 Effects of vacuum polarization in S-wave proton-proton scattering, j.P. Naisse, page 873, Bulletin Academie Royale de Belgique, 1972.

electrostatic energy being randomly modified by the creation of virtual **(e+, e-)** pairs and possibly disturbing, if not destabilising, the biological environment, by triggering harmful chain reactions via free radicals.

In thermodynamics (Landau, 1994) specified that the number of particles also depends on the conditions of thermal equilibrium, he assimilated pair creation (Carl Anderson, 1932) to a chemical reaction while considering its relativistic expression.

-   For a temperature of $T \sim mc^2$ the number of pairs would be very high, and corresponds to the Fermi-Dirac statistic, which takes into account the indistinguishable quantum effects in the distribution of fermion energy states (protons, electrons) at thermodynamic equilibrium.

-   For a temperature $T > mc^2$ the number of pairs would correspond to *0,183(T/ħc)V* and an energy, $E^+ = E^- = ((7\pi T^4/120(\hbar c)^3))V.$

-   For a temperature $T < mc^2$ the number of pairs would be small, i.e. *exp (-mc² /T)*, but not negligible at the biological temperatures of homeotherms, particularly heterotherms, during low-temperature hibernation.

The threshold energy required to create a pair is :

$$2m_e(1+ m_ec^2/m_Rc^2) \qquad\qquad [\,1.5\,]$$

If me is the mass of the electron and mR is the rest mass of the protons, the ratio $meC2/mRC2$ is the inverse of the fine structure constant *a.*

New questions arise: what impact does the production of pairs *(e+, e-)* between $T < mc^2$ et $T \sim mc^2$ à $T > mc^2$ have on the electronic balance of atoms and ions, on the rejection of electrons and protons, on the production of free radicals and their chain reaction in the mitochondria and the cell?

The proton, dissociated from the electron, is in an excited state and loses energy in matter. This linear transfer with the atoms slows down the proton, which is in search of an electron. It can be associated with a new chemical bond by a receiving atom, resulting in a protonated molecule within a complex protein whose variation in arrangement is likely to trans-localise it. Under this last term, there are imperatively two aspects to consider, on the one hand a Schrodinger-type wave, whose non-localised binding energy is representative of the state of the excited proton subjected to the potential of a flow of electrons that characterises the proton-motive force, and on the other hand, the strictly biochemical aspect that arises from its temporary binding to the host site of its acceptor, which effectively trans-localises it.

In comparison, electrons are transported on the valence layers of atoms during successive oxidation-reduction processes, so the more appropriate term of translocation for the proton suggests another form of displacement, because the $H+$ proton ion is extremely reactive. In a cell, the proton hydrate is translocated by protein complexes inserted into the inner membrane of the mitochondria.

The notion of a quantum field associated with Schrodinger's equation seems more appropriate for interpreting its path channelled by an arrangement of proteins whose atoms are as many obstacles to be crossed. By the tunnel effect, the probability of a proton crossing this potential barrier is non-zero, as the movement of a non-localised stream of electrons, resulting from oxidation-reduction, interacts with the mobility of the protons.

"The proton is used for its resonance capabilities in
magnetic resonance imaging and protontherapy,

irradiate a tumour with great precision and at a finely-targeted minimum therapeutic dose".

Returning to our hypothesis, it would be a change in the state of metastable atoms and molecules subject to quadrupole interactions caused by a disturbance originating in the inter-membrane space of the mitochondria that would trigger the coordination of cellular activity by means of a quantum field of pulsed protons, at saturation and at threshold. The celestial nature of this field of *H+* protons and its virtual photons is necessarily greater than that of interactions between molecules. Does this statement justify biochemist Teresa Gordon's proposal that this field should act as a coordinating unit for cellular activity?

After these preliminary considerations about the proton, let's make the transition from quantum physics to statistical physics. According to Lev Landau (1994), the negative temperatures we are exploring are a property of dielectric and paramagnetic atoms whose magnetic moments orientate quite freely, and their interactions induce a magnetic spectrum that corresponds to all combinations of magnetic moment orientations.

This new state of interacting moments manifests itself in positive and negative temperatures between :

$$T \ < + \, 0, + \infty, - \infty, - \, 0 >$$

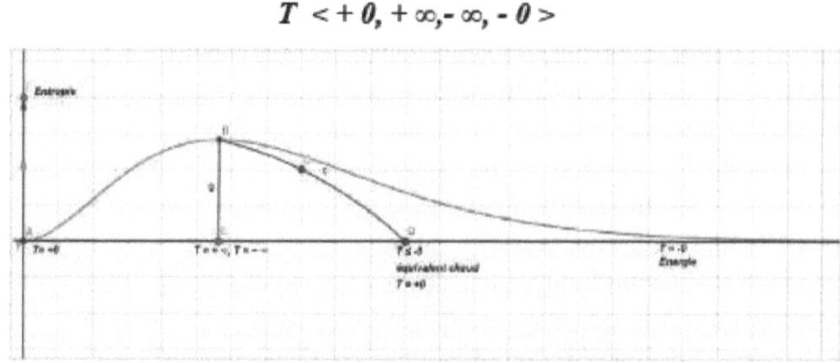

18. Graph representing this distribution,
of the form[16] *(x/ex-1)* on the x-axis are
the time intervals *T* and the increasing energy *E*,
and on the y-axis the entropy *S*.

(i)   For a temperature equal to + *0*, the interacting atoms are in *a "low energy"* quantum state with zero entropy.

(ii) When the temperature increases from *+ 0 à + ∞,* , relative to the number of interacting atoms, to *(+∞,-∞)* the average energy and entropy are at their maximum, and identical.

(iii) From *- ∞ à - 0,* when energy increases, the initially high entropy decreases and tends to cancel out for a temperature at - *0,* when energy increases.

The graph is characterised by an asymmetry in favour of the cold (iii), for a temperature below minus zero which would present *"a slight excess of entropy for an energy*$^{\Delta\varepsilon}$ *which tends to rise"*, compared with temperatures above zero, initially with low thermodynamic energy (i).

For a tempĕrature tending towards - *0* compared with the positive tempĕratures*(+ 0, + ∞),* due

---

16 According to the Poisson distribution of a black body.

to this cold asymmetry, there would be a greater ëcart of the distribution of atoms in the cold. When entropy initially ëlevëe, dëcroit and tends to cancel out, the ëtat quantum would be stronger when energy increases in this space which would effectively merge with time.

Since negative temperatures ëare higher than positive ones, at a temperature of - *0* to a few negative degrees $_{T\text{-}o,}$ this difference would present a small potential energy $^{(\Delta\varepsilon)}$ and a "*momentarily*" non-negligible entropy:

This $^{\Delta\varepsilon}$ of energy and entropy, localized towards point *c of* the graph, would be intrinsic to the binding energy of the atom, which we evoked earlier, the random distribution of the charge of the proton intervening in the free energy of an atom compared to its spin. This would confirm the significant difference in entropy and energy ($^{\Delta\varepsilon}$), when the latter increases at very low temperatures between $_{T\text{-}o}$ *and -0 °C.*

Thus, for a temperature $T = _{T\text{-}o} < - \mathbf{0}$, the heat capacity would be *2.80 N* $_{atoms/T\text{-}0}$, for $_{T\text{-}0}$ between - 0.1°C and about - 15°C, at most in the case of extreme hibernation, is this enough to obtain a state close to that of a degenerate Bose gas?

This demonstration gives us a glimpse of a fleeting space of biological resolution where there would be :

"A low energy
that increases in excess as entropy decreases".

A singular situation at very low temperatures, if we consider that at 0°C the mean square velocity of hydrogen is 1.85 ms^{-1} .

In the case of a crystal subjected to a strong magnetic field that reverses, causing a rearrangement delay of the spins, if the magnetic field is removed, the temperature effectively becomes negative, which for convenience we refer to as an exclusively quantum "*cold solution*". After a relaxation time, the temperature will equalise in the thermodynamic range of a "*Warm Solution*" through the exchange of energy between the spins and the paramagnetic crystal lattice, which is no longer disturbed by a magnetic field.

The Earth's magnetic field has an average value of around 50 micro-Tesla, or 1/2 gauss for the Earth, and has undergone pole reversals and fluctuations in intensity over the geological eras. The magnetic field measured by the Swarm probe (2014) gives values that vary between 20,000 and 60,000 nanotesla, increasing from the equator towards the poles. The Earth's magnetic field is thought to be of the order of 30 to 70 micro-Tesla for an electric field of 150 v/m.

In the time and space of the species revolution, several natural selective constraints need to be brought together. The drifting of the continents and the formation of the oceans are associated with inversions of the Earth's magnetic field. We need to consider the consequences for the intensity of radiation due to the penetration of cosmic protons, depending on the solar cycles that cause electromagnetic disturbances. These cosmic and telluric events are to be correlated with the climate fluctuations on which the biological revolution of species, that of primates, the human species and the transformations of civilisation depended and still depend. This is particularly true of glacial and interglacial periods, which are very hard on animals and human beings, who are forced to hibernate in cold periods or adopt individual and collective survival strategies.

30

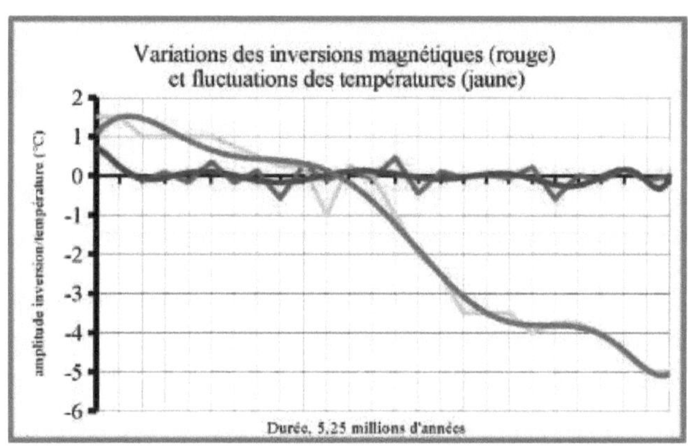

Millions of years from the end of the Tertiary to the Quaternary.	Recent reversals	Polarities (+ normal/- inverse) and duration of magnetic inversions	Temperature (°C) global average marine sediments	Paleoclimat Evolution of plant and animal species	Evolution of Primates and Hominins	Transformations in Civilisation
5,25	Thevra		1,5	Pliocene	Proconsul Dryopithecus Orrorin tugenensis	Divergence of Panines and Hominines
5,01	Thevra	0,24	1,5	Global cooling	Ardipithecus ramidus	Use of stones, sponges, rods and levers.
4,89	reverse	-0,12	1	Mountain and river formation		
4,81	Sidufjall	0,08	1	Flora of temperate climates develops		
4,64	reverse	-0,17	1	Conifers in the North Meadows to the south		
4,47	Nunivak	0,35	1	Presence of mammals, disappearance of the largest ones	Austalopithecus afarensis/africanus barhelghazali	Percussion tools?
4,29	reverse	-0,18	0,75			
4,17	Cochiti	0,12	0,5			
3,59	reverse	-0,58	0,25			Tool marks
3,33	Mammoth	0,26	0,25			Fitted pebbles
3,22	reverse	-0,11	-1			
3,12	Mammoth	0,1	0,25			
3,05	reverse	-0,07	0		Homo habilis	Use of splinters, cleavers and scrapers
2,59	Kaena	0,46	-1			
2,14	Inverse	-0,45	-2	Pleistocene		
2,13	Meeting	0,1	-2,5	Cool climate		
2,08	reverse	-0,05	-3,5	Arctic flora		
2,06	Meeting	0,02	-3,5	Rhinoceros and woolly mammoth		
2	Inverse	-0,06	-3,5		Homo ergaster Homo erectus	
1,78	Olduvai	0,22	-4			Biface cutting
1,19	Inverse	-0,59	-3,75			
1,18	Cobb Mountain	0,01	-3,75			
1,06	reverse	-0,12	-4			
0,9	Jaramillo	0,16	-4,5	Glacial and inter-glacial periods Gunz Mindel Riss		Traces of fireplaces, chips, retouching, blades and tips.
0,78	Inverse	-0,12	-5	Variations in sea levels		Abbevillien Acheuleen Levallois
0	Brunhes	0	-5	wurm	Homo sapiens/ Neanderthalensis expansion towards Asia and the whole world	Mousterian Micoquian blades, pompons, propellers, needles, chisels. Ornate caves

19. Diagram and table of terrestrial magnetic inversion, from the end of the Tertiary Era (Pliocene) to the Quaternary (Middle Pleistocene), correlated with climatic events and temperature fluctuations, with regard to the revolution of plant and animal species, Primates, Hominins at the beginning of Civilisation, by reversal effect.

As a naturalist in biological evolution and

In order to understand the transformations of civilisation, I have concentrated on the alternation of cold and warm climatic conditions during ice ages. In particular, on the phases of natural hibernation at altitude or artificial expérimental which optimise observations of the cellular environment at low températre. At this level of resolution, my aim was to modëlise the effects of a quantum field of excited protons and the modalities of their translocation in

31

protein complexes embedded in the inner membrane of the mitochondrion.

These relationships between temperatures and magnetism force us to introduce thermodynamic considerations linked to the nuclear spin, as a fundamentally random parameter. We have described the breakdown of fine structures into hyperfine structures, whose energy intervals are small, but not negligible with respect to temperature. These hyperfine structures are most likely to appear at temperatures of between 0.1°C and 1.5°C, close to the low biological energy hibernation state$^{(\approx \Delta\varepsilon)}$ of animal species. The splitting into hyperfine levels would depend on a random nuclear term that includes the spin *(i)* in the *nuclear* free energy *(F)* resulting from its product with the number of atoms *(N)* in correlation with the temperature *(T)*, i.e. according to Landau :

F nuclear = - NT In (2i+1) for a nuclear spin greater than zero, *i = +1/2* of the proton   [1.6].

In relation to spin, the entropy $^{(\varsigma \cdot}$ *nucleaire)* and the chemical constant *(Z nucleaire)* are well modifiëes by the nuclear term *In (2i+1)*, indëpendently of tempërature, in particular when *F nucleaire* corresponds to an energy variation$^{\Delta\varepsilon}$ equal to *T-o*, so :

If *T-o < - 0 °C*,

$$F\,nuclear = - N\,(\Delta\varepsilon = T_{-0} < - 0\ °C)\,ln\,(2i+1) \qquad [\,1.7\,]$$

The domain of this "*cold solution*" would be more quantum than thermodynamic, in these conditions, we postulate the following:

S nucleaire = N In (2i+1), the entropy which depends on the number of atoms tends to increase, and to decrease for [1.7]

Z nucleaire = In (2i+1), the chemical potential would be sensitive to the diamagnetic, paramagnetic and random ferromagnetic susceptibility of the spins independently of the temperature and the number of atoms.

Subsequently in the models, we consider that the average fluctuation of atoms in the mitochondrial and cellular environment varies from their initial state undisturbed by a field. For the model, we generalise two simplified forms of pseudo-liquid crystals characterised by changes in fluidity and conductivity as a function of the arrangements and rearrangements of the atoms and molecules, whose metastability is linked to their susceptibility to a perturbing electromagnetic field, i.e. :

-     $1/\sqrt{N}$ for the inner and outer membranes, which can be likened to a nematic liquid pseudocrystal arranged in layers.

-     *In)* $1/\sqrt{N}$: for protein complexes with a cholesteric appearance included in the inner membrane.

(i)   Their very low negative constant ditimagnetic susceptibilityv would give a spontaneous magnetisation inversely proportional to the field.

(ii) Their paramagnetic susceptibility, a low positive constant, would be proportional to the field.

(iii) Ferromagnetism presents a cycle of hysteresis, saturation and magnetic remanence.

The metal ions present in the cellular medium, as oscillators, will react under the influence of field variations and rearrange themselves according to their magnetic susceptibility, in particular :

(iv) The alkalis $Na+$, $K+$ and Cl^- transport electrons, which are particularly condensable.

(v)   The alkaline earths, Mg^{2+} and especially Ca^{2+} , are discussed below.

(vi) Highly catalytic bio-activator metals with paramagnetic properties: Fe^{2+} , Co^{2+} , Ni^{2+} , Cu^{2+} or diamagnetic Cu^{+} , Zn^{2+} , reversible oscillators in the cell environment.

(vii)In particular, the *H ion*, which has great ionic mobility thanks to its own mass, its kinetic energy and its magnetic potential, with its properties of kelicity and translocation, which effectively condition the production of vital energy in protein complexes.

This energy is used by the cell where the hydrogen atom has a weak link function, connecting the bases of the DNA double helix, which diverges during its duplication. The two strands are sëparësed using the energy of ATP degradation, during replication, which precedes transcription into RNA, before translation into specific proteins.

In the mitochondrial intra-membrane space, the hydrogen ions, accelerated by the flow of electrons from the protein complexes, which are excited $H+$ to highly excited $H+*$, acquire a high effective cross-section.

Interference from *s-waves* associated with the field of highly excited $H+*$ protons, in strong interaction at saturation with atoms, molecules and proteins in the mitochondria and the cell, would increase its *a-scattering* length, a significant criterion for assessing the propagation modes of the threshold pulse wave and its divergent derivatives.

The atoms in the mitochondria and in the cell environment are paired oscillators, whose mean fluctuations in the membranes modelled in $1/\sqrt{N}$ and in the protein complexes in $ln(\ )$ $1/\sqrt{N}\ ^2$ are likely to modify their configuration under the effect of this field of excited $H+$ protons, to very excited $H+*$. This field would change spontaneously from weak to strong in the mitochondrial intra-membrane space, depending on the variation in its acidity $(\Delta\,pH),$ perturbative à saturation, on which depends the frequency of the pulse field resulting from its own quantum interactions, and proportional at threshold, to the differential magnetic susceptibilities of the atoms that make up the membrane and the protein complexes.

In our forward-looking approach, we have opted for two approaches
solutions :

-   A *"Cold Solution"*, if the frequency of the wave is greater than the tempërature, i.e. a *nuclear* free ëenergy $F,$ equivalent to an energy *As* greater than $T_{-0} \le -\,0\ ^{\circ}C$. In this case, the distribution would be purely quantum, discrete and fundamentally random, with a probability of divergence of the resulting waves that would occur at very high speeds between the highly excited $H+$ particles at saturation. At threshold, the unitary spinor field resulting from interactions between the highly excited $H+*$ and the perturbed inner membrane atoms ($1/\sqrt{N}$ and $ln(\ )$ $1/\sqrt{N}\ ^2$, would be likely to generate several quantum energy states depending on the spin orientation with probable states, assumed to be close to Bose-Einstein condensation. Biochemical rearrangements would occur, assuming a non-negligible residual entropy and relatively high energy, in a discrete quantum state that is more remarkable at low temperatures.

-   If, on the other hand, the frequency of a *"hot solution"* becomes lower than the temperature, there is no longer any quantum distribution to consider, and we reach the thermo-hydro-dynamic domain of discernible fluctuations-dissipations that would occur after the indiscernible cold solution, depending on the calorific capacity of the organism in question, whether homeothermic or heterothermic, which is more sensitive to low temperatures.

These hypothetical considerations led us to look for cases of animal species that survive at low temperatures, at altitude, and that practice hibernation, such as amphibians (Anurans and Urodeles) and reptiles, in order to study and demonstrate, during the transition from a weak to a strong field, discrete quantum effects and their possible consequences, particularly on the coordination of cell activity.

- *In the case of the weak field,* the magnetic susceptibility of atoms is distinguished by a paramagnetic term that depends on the magnetic moment of the electrons in the atoms and a diamagnetic term that corresponds to the orbital motion of the electrons. In this case, the susceptibility is considered to be essentially paramagnetic.

- *When the field becomes stronger*, the resulting wave of the pulsed proton field, at threshold, would produce *a* high-frequency *"oscillation"* that would initially not depend on temperature.

This oscillation, which would be strictly quantum, would grow exponentially as a result of its entanglements and would normally become negligible after its thermo-hydro-dynamic absorption by fluctuations-dissipations in the biological pseudo-crystalline medium maintained in an average metastable state, at an optimum temperature, for the cell and the vital coherence of the animal.

As a result, initially independent of temperature and then as it increases, entropy being by definition an extensive quantity, the number of microstates increases, as do their capacities for arrangement and rearrangement as a function of their biochemical potential *(Z)*. This combinatorial process takes place at different levels of emergent integration, notably during cell division and the development of an organism, metamorphosis or organ regeneration. This structuring and/or destructuring branching usually remains under genetic and epigenetic control, in correlation with the integration level units (proteins, cells, animal) subject to the sëlective constraints of the environment. Nevertheless, we have to reckon with the interferences resulting from the differential interactions of the quantum field of protons and photons at saturation, pulse at threshold, with the atoms of the protein complexes and their supporting membrane, generating, during their rearrangement, what I call :

"Divergent wave derivatives of cold solutions

Indeed, they are reactive in quantum domains through indiscernible incidental divergences, which moreover, as a function of their diffusion length, would spread out in space/time, initially independent of temperature correlatively to the discernible thermo-hydro-dynamic hot solutions that follow them, masking the quantum effects...

Let us continue our exploratory hypothesis in an environment considered to be firm, such as the intermembrane space of the mitochondrion during the cell's long metabolic phase followed by a brief period of activity. Let's consider the action of the field *(H+)* of N_{p+} excited protons H_{+*} with an average magnetic moment $^{(\mathcal{M})}$, during the change from weak acid *(pH)* to strong acid *(pH*)*, i.e. a variation in the field of excited protons $^{(\partial\ H_{+*})}$ which depends :

- The coordinates of protons, i.e. their quadrimoment, i.e. three coordinates in space conjugated with one in time, assuming that their quantum dimension has a discrete and random resolution corresponding to the non-homogeneous charge distribution of the proton, mentioned earlier.

- From their pulses to the saturation threshold, i.e. **a,** their length
diffusion increases with acidity, from low to high.

- Of their spins, either *ln (2i+1)*, a random nuclear term which conditions the free energy and its relation to tempërature. In particular its nuclear free energy, *F nucleaire* for a$^{\Delta\varepsilon}$ supërieur than $\overline{T_{-0}} \leq - 0\ ^{\circ}C$, siëge of the cold solutions of the discrete quantum interactions, nevertheless structuring or destructuring.

Under the mutual and increasing repulsive influence of the highly excited protons H^* as a function of the differential reactivity of the atoms of the more or less polarised medium of the internal membranes and of its protein complex components *(Ho)*, its variation in energy at saturation threshold *(dH+*)* by the average magnetic moment of the protons$^{(\mathcal{M})}$, expressed by its Hamiltonian is :

$$\partial\hat{H}s = -\,\mathcal{M}\,\partial H+^* \qquad\qquad [\,1.8\,]$$

According to the variation *dHs* of the Hamiltonian of the pulse field has threshold $H+^*$ of $_{Np+}$ protons, the Hamiltonian of the atoms of the mitochondrial membranes and of the cellular medium becomes :

$$\hat{H}_{milieu\ cellule} =$$

$$\hat{H}_0 - \partial\hat{H}s + ((1/\sqrt{N})\ (e^2/8mc^2) + (ln(1/\sqrt{N})^2(e^2/8mc^2))\ (Hr)^2 \quad [\,1.9\,]$$

The perturbation of the atoms' electrons in the form of nematic *(*$^{1/\sqrt{N}}$ *)* and cholesteric *(ln(* $1/\sqrt{N}$ *²*) liquid pseudocrystal is caused by the first two terms$(\hat{H}_0 - \partial\hat{H}s)$ of the field predominating over the thirdiëme and fourth terms. According to$^{\Delta\varepsilon\ \sim\ \partial\hat{H}s,}$ their resulting energy levels, the spins of the atoms orientate themselves according to the pulse field that has reached its saturation threshold, which affects their respective susceptibility through a transition from their fine structures to ultrafine structures. The new atomic configuration of the molecules would disrupt the biochemical conformation of the membranes and protein complexes, affecting their conductivity, fluidity and permeability, which would trigger the unitary exchange activity of the cell with the extra-cellular environment. Is this coordination

of individual cells sufficient to induce the animal's vital coherence? When the energy level$^{\Delta\varepsilon}$ of the hyperfine structures of protons and atoms in strong interaction is greater than the temperature, the susceptibility of atoms and molecules would be of a strictly quantum nature, and this is the area we are exploring at low temperatures. Indeed, when we go down to

temperatures between - 0.5°C and -0°C has an energy$^{\Delta\varepsilon}$ greater than the cold temperature $_{T-o}$ which tends, from -0 to $^{-\infty,}$, to a limit temperature equivalent to the hot temperature $_{T-ec}$. A small, non-negligible potential for entropy excess and a sufficiently high energy, effectively independent of temperature, would indeed exist in an exclusively quantum resolution space that we still need to study in greater depth in order to demonstrate probable underlying effects. If the atoms are diamagnetic, the effects are masked by paramagnetism, as susceptibility does not depend on temperature: the oscillation term is not insignificant and its structuring and/or destructive consequences remain to be considered in the case of induced superconductivity and divergence of the derived wave. This could be an aleatory effect that I have demonstrated during the transition from linear to non-linear in complex evolutionary processes, which takes place for very low values of the adaptability coefficient (around two) and during intermittencies with very singular structures and great variability.

On the other hand, if atoms are paramagnetic, their susceptibility is inverse to temperature, according to Curie's law. The paramagnetism of ions and free radicals would follow Langevin Curie's law, which has applications in the field of very low temperatures.

When a magnetic field is applied to a paramagnetic substance such as a bio-activator metal, there is a uniform distribution of magnetic moments, initially distributed in all directions, as they align with the field, energy is removed in proportion to the strength of the magnetic field. However, as a result of thermal collisions and vibrations, *"the temperature that had been lowered"* will increase through the magneto-caloric effect, blurring the cold solution, which is difficult to discern.

To sum up, there are several regimes to consider, that of the oscillations of the field of protons and photons at saturation, pulse at threshold, with its very problematic *"cold, strictly quantum solutions"* relating to the quantum entanglements which provoke divergent derivative waves whose normal and abnormal, if not naturally pathological, effects must be analysed and understood. In particular, by going back to the initial conditions of *'divergence'* of the chain reactions involved in the functioning of the cell, including the *'thermo-hydro-dynamic hot solutions'* of the fluctuations-dissipations that follow on from the biochemical interactions observed in the model approach below.

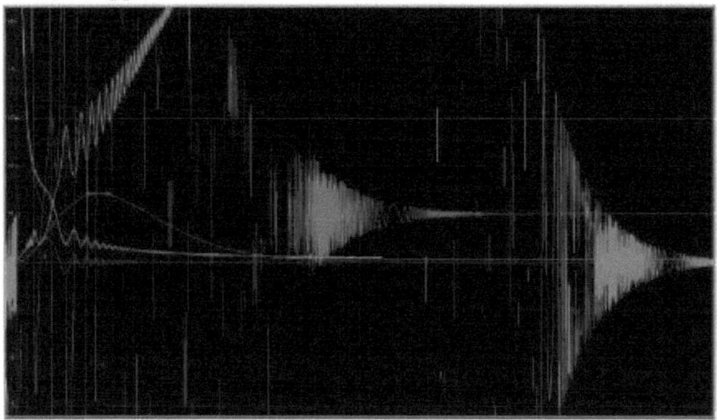

20. Model approach, simulation of a pи^ё field of protons at
saturation and threshold in 1 mitochondrial intermembrane space.

-    The light red curve with the green arrow indicates the transition from a low pH to a high pH of an intermembrane field of very strong protons, whose power wave at saturation of protons and photons is represented in blue.

-    This wave reaches a threshold of mutual interaction with the membrane atoms (dark red), and the resulting variations in the spinor field (light blue) cause their biochemical rearrangement.

-    We identify a potential well (thin green, left) which could indicate a superfluid^ and superconductiv^ consëcutive to cold ëphëmëres solutions, could it be a pseudo Bose-Eintein condensate?

-    Quantum entanglement and interference are simulated by a transformëe of the perturbation, its first and second dërivëes (local max) reveal the divergence of the dërivëes waves (in
green, right).

- Following the indistinguishable cold solution, the hot solution grows exponentially through fluctuations and dissipations in the thermos-hydro-dynamic domain constrained by the activity of the cell in its biological environment (homëostasis) and of the animal in its environment, during its adaptive ëvolution.

The incidental variation in the probability density of the proton and photon field in relation to the values of the saturated pulses, which have reached a threshold compatible with vital coherence, could play a role in shifting the critical transition from linearity to higher non-linearity at higher values, which would validate the stochasticity of adaptability, whose term of variation is considered to be fundamentally random.

As is often the case when trying to find a solution to a complex problem, I use a hypothetical parallel path. Here I use the non-linear Schrodinger equation of Eugene Gross and Lev Pitaevskii, which characterises the state and dynamics of a perfect gas of ultrafast bosons independently of time. This equation seems applicable to the saturation pulse field of protons and photons excited to very excited in the inter-membrane medium, which is closed for a long time during the cell's metabolic phase. Spontaneously open in the mitochondria during cellular activity, as a wave function of bosons[17] resulting from the interaction of $_{H+*}$ protons with each other and with the excited *"trap"* atoms in the mitochondrial and cellular environment. Proportionally to the variation in *pH*, which would go from a weak to an excessively strong field, provoking a brief high-frequency oscillation that could be thresholded at low tempërature between *-0* and $_{T\text{-}o}$ when the tempërature tends towards - $_{Tec}$

ëquivalent hot, at an energy close to $\Delta\varepsilon$ strictly quantum, initially without any prëpondërant thermodynamic effect. This led us to consider underlying, ateatory and fleeting quantum effects, which we dësignë trivially as a probable discrete *"Cold Solution"*, masked by the fluctuations-dissipations of the proton/photon and atom interaction wave in the thermos-hydro-dynamic domain in the mitochondrial and cellular milieu, during *"Hot Solutions"*. Without going into detail, it seems interesting to adapt this equation to our test model.

$$\mu\, \varPhi\,(r) = (- h^2/2\,m\, {}^{*}\nabla^2) + V_{ext}(r) + \Delta N g\,|\,\varPhi\,(r)\,|^2)\, \varPhi\,(r) \qquad [\,1.10\,]$$

. · ΔN bosons in the field of exc^sed protons *(H+*)* *in* relation to the variation in *pH* from low to high in the intermembrane space.

· $\varPhi$ its wave function.

. *m* mass of the particle (proton).

. *g* constant dëpendent on the scattering length *a* of the interaction potential between bosons with zero energy.

. V_{ext} confinement potential of the atoms in the intra-mitochondrial space.

· μ chemical potential, or $\zeta_{nuc} = ln\ (2i+1)$ such that $_{Snuc} = N\ ln\ (2i+1)$ indëpendent on tempërature, has a decreasing entropy ëlevëe for an energy ë that tends to increase at low tempërature.

The questions that arise, are to know, how ëvolve the divergences of cold solutions in the long time

of quantum resolution space?

We have dëcribed the proton which, along with the neutron and electrons, are fermions of

---

17 If the bosons are considered in the same Bose-Einstein state at very low temperature, the Hamiltonian satisfies the following normalization condition:
$\int |\Psi|2\, d\,V = 1.$

spin 1/2, according to Pauli's exclusion principle, they cannot be in the same ë quantum state. Whereas bosons like photons, arising from their interactions in the form of a quantum field, would tend to clump together in a single quantum ëstate, including in a composite assembly of bosons and fermions.

This condition would appear to be met in the inter-membrane space of the mitochondrion, where the protons, highly excited due to their repulsive force arising from their non-homogeneous charges, as they clump together, emit photons. The resulting composite wave of saturated and pulsed protons and photons would be trapped by the atoms of the inner and outer membranes, which at threshold, depending on their magnetic susceptibilities, would trigger the short phase of activity and the waves resulting from these complex interactions, more or less divergent, which raise questions about their long-term resolutions.

This electromagnetic trap would be sensitive to the increase in the number of over-excited protons at high celeration, in particular through the production of a presumed "*cold*" quantum expression as opposed to the "*hot*" one, which would occur in the resolution space when the temperature tends from *-0 to -m*, when the *As* energy is greater than the temperature between -*0°C* and T_{-o}, when T_{-o} tends towards $-Tec$, compared with the hot equivalent.

Under these initial conditions, the ephemeral and indistinguishable "*cold*" quantum effects would precede the fluctuations-dissipations of the wave in the mitochondrial and cellular milieu which normally lead to the "*warm*" and discernible thermos-hydro-dynamic domain, which stabilises at around 37°C in the organism of a homeothermic animal compared with a cold-blooded, hëtërothermic organism, close to the outside tempërature. Particularly during hibernation at very low temperaturërature, which could approach a critical temperaturërature (*Tc*) conducive to a state of condensation close to that of Bose-Einstein, which remains to be demonstrated.

Each proton, according to the Pauli exclusion principle, would superpose itself in different quantum states in this piëge of intermembrane confinement. Under these conditions, could it adopt a degenerescence regime when it reaches a critical temperature (*Tc*) of the medium that is lower than the Fermi[18] temperature (T_F)?

When the temperature of the medium is lower than the Bose-Einstein condensation temperature, the photons produced by the weak to very strong interactions of protons confined in quantum interference with the atoms of the inter-membrane trap tend to condense in the fundamental state of the composite trap. Here, in a brief, discrete, indistinguishable "*cold*" quantum degeneracy. Therefore, by applying the formulae below, two solutions could be envisaged at low temperature, for *N excited* protons, we consider *N excited* photons at the following energies.

- Photon energy :

$$k_B Tc = h\omega \ (\ 0{,}83 \ N \ excités \ )^{⅓}$$
[1.11]

- The energy of protons :

$$k_B T_F = h\omega \ (\ 6 \ N \ excités \ )^{⅓}$$
[1.12]

Let k_B be Boltzmann's constant, as a function of the oscillator frequency **w** of photons or protons, and critical tempërature *Tc* or Fermi tempërature T_F respectively. Let's look at how ëvolve this supposed гайёге condensëe communement dësignëe "*molasses*" a dëfaut to name it Bose-Einstein condensate, which would be premature. At maximum impulse, the protons

---

would be in an *N excited* state, which becomes smaller than *N excited* photons of interactions between the protons and the atoms of the piëge medium, the field of quantum interactions of the photons would reach saturation, according to :

$$N\ saturation = 1,202\ (k_BT / \hbar\omega)^3 \qquad [\,1.13\,]$$

As a result, if the temperature decreases to a critical temperature *Tc*, between *-0°C* and T_{-0} close to - T_{ec} which would tend towards - ∂a / $+\partial a$, the ensemble would find itself in a state close to Bose-Einstein condensation. At least, in a certain proportion at a condensed minimum, for a temperature lower than :

-0°C pour un + Δε, à un T₋₀ proche de T critique des photons *>T*F des protons.

This state manifests itself as a very sharp peak of condensed photons whose wavelength corresponds to the fundamental state of a potential well, surrounded like a Mexican hat by the uncondensed excited photons and protons at the broader wavelength. In this prospective research hvpotlK'se, we would go from a disordered state of excited photons to a more homogëne wave formation when the temperature decreases below the Bose Einstein critical temperature, in correspondence with the de, de Broglie wave. Degenerescence would occur when the de, Broglie wave approaches the distance between protons at very low temperatures, in a state that can be described as, de, Broglie.

bio-plasma.

For *H+* , the proton is considered:

- of low mass, the de Broglie wave varying as a function of *m-72.*
- gaseous at all tempëratures.
- easy to polarise in a magnetic field.

The initially weak interactions become strong, to very strong, with the consëquence of gënërer a differential threshold at saturation of protons and photons of strong interactions in relation to the atoms of the intermembrane trap. The consequence of this would be the spontaneous and indiscernible acquisition of a probable superfluidity and superconductivity of the membranes by acting on the rearrangement of the protein complexes which would find themselves out of unstable equilibrium, linked to a hypothetical pseudo-condensation of the field of protons and photons at saturation. A cold solution whose relaxation time with the warm solution would balance the organism's temperature, enough to ensure the survival of an animal in a state of extreme hibernation, for example in the Arctic frog *Lithobates sylvaticus*, which survives frozen at -16°C by increasing its uremia and glycemia during cryo-protection. The attached appendix is devoted to the Greenland shark, which lives at very low temperatures at great depths.

21. An amphibian observed hibernating in a stream on Mont Lozere, at 1400 metres.

This state of survival remains to be explored, the detection of condensation is difficult, the ephemeral condensate is a very unstable state, however, it should be remembered that the alkalis mentioned above, which are very present in the cellular medium (sodium, potassium) are condensable.[19] At short distances, their interaction potential is highly attractive, resulting in numerous linked molecular states. It is the amplitude and length of diffusion *a* that characterise the effects of long-distance interactions between protons and atoms on the physical properties of the condensate, i.e. *at* the first zero of the wave's sine wave, when it cancels out.

- When *a* is greater than zëro, the energy is cinetic $(Eo \wedge Emax \wedge Eo)$, the wave would reach a maximum narrow peak, very punctual, between two probable limits of condensation, located between the inflection points in the modële. In our configuration, the **nuclear** free energy *F* would be low, of the order of $E_o \wedge As = Emax$, but affected by a sufficiently high entropy for a very short time which would nevertheless be significant, if not predominant, at a very low temperature below - *0*.

- When *a* becomes less than zero, the wave collapses into its potential well, into the resonant piëge of protons and photons that are momentarily very repulsive, then attractive, due to interactions with the perturbed trap atoms, which spontaneously rearrange themselves, so that the trap does not remain sufficiently attractive to maintain the pseudo-condensate. The energy $Eo \wedge As \wedge Eo$ would increase towards a classical thermodynamic regime of fluctuations-dissipations by reducing entropy until a low limit temperature of biological survival between - *T limite de survie* and -*0,* corresponding to the integration of *As*.

In the present case, according to the simulation of the approach modële, the resolution of the pulsed field, saturated at threshold, of protons and trapped photons interacting with atoms, presents a singular peak, at the maximum of the potential well, so the relative context *"after"* a strong repulsion would be momentarily and sufficiently attractive to produce a very short-lived pseudo-condensation. Having put forward this hypothetical proposal on the genesis of a biological condensed state or bio-plasma, I will end this chapter with some comments on oscillations and quantum chaos experimented on hydrogen. Martin Charles Gutzwiller (1925-2014) described chaos in classical and quantum mechanics (1990).

---

19 Comparatively, the melting temperature is -250.16°C for hydrogen, 44.15°C at 590°C for phosphorus and 113.7°C for iodine, a key biological constituent along with the alkalis, potassium: 63.5°C, calcium: 842°C and sodium: 97.8°C.

Dans un espace fermë, tel l'espace intermenbranaire, l'onde *y* envahit la cavite, ce qui provoque des interferences complexes lors du passage d'un regime regulier semi-classique à un probable regime chaotique lie d'une part, to the dynamics of the proton field pulse to threshold in a proton-photon association as we have just described, and to the electrons of the atoms (non-localised) with unstable periodic orbits (redox electron flow, electron ejection, pair creation and free radical formation). If the observed energy levels cross in a regular regime with exponential signal decay, the random repulsions of H^{+*} would be accentuated towards a divergent non-linear regime, by the incident spacing of the energy levels, a probability normalised from a Poisson curve in the diagram above. As a result of these interactions, the divergent drift of the pulse wave altered by random interferences from the cellular environment could trigger a natural pseudo-quantum chaos, whether *normal or incidentally forced*, with the risk of being more destructive than structuring through deleterious chain reactions, at the level of the cell and the animal, in Homo sapiens in particular.

In this revealing chapter, we have laid the foundations for the transition from quantum mechanics to statistical physics to demonstrate the unitary action of a threshold-pulsed proton field in the intermembrane space of a mitochondrion, as :

Hypothesis 1: A coordinating unit for cell activity, based on a proposal by biochemist Teresa Cordon.

Hypothesis 2: The internal temperature is regulated by an equilibrium of *"indistinguishably cold"* quantum solutions with a pseudo-condensed tendency, and subsequently, *"indistinguishably hot"* by fluctuation-dissipation in the thermo-hydro-dynamic domain.

Hypothesis 3: On the other hand, the divergences of the initial quantum field and its dërivëes would have an aleatory transition probability towards non-linearity, the consequences of which remain to be explored in the long time of their biological resolution space.

To support our proposals, we need to describe the mitochondria in detail before proceeding with test modelling, in order to validate the fundamental role of the threshold pulse proton field as a coordinating unit of cellular activity.

Next, we need to evaluate these hypotheses on the maintenance of the homeostasis of integration level units and estimate their consequences on adaptability during species revolution. Are these critical conditions outside the highly unstable equilibria of chaotic divergences likely to cause disorders by disrupting the vital coherences conditioned by the transport of electrons in conjunction with the translocation of protons in mitochondrial protein complexes?

Chapter 2

Mitochondria and basic proteins.

Cold-blooded species and those that hibernate are confronted with a drop in their body tempërature, in order to function at low biological ëergy, in these drastic conditions, the composition of the cell membranes and mitochondria are modified by adjusting their fluidity, permëabilitë and conductivity. The complexes of protëines included in these membranes that contribute to cellular respiration are also affected during states of thermal and hydric stress as a function of disturbances in the cellular and extra-cellular milieu when the organism reacts to climatic fluctuations in the environment.

We pointed out in the previous chapter that atoms obeyed Pauli's exclusion principle by quantising their energy levels, while at low temperatures they would find themselves in a single quantum state in the form of a probable Bose-Einstein condensate. A sort of ephemeral bio-plasma, conducive to superfluidity and superconductivity at a critical temperature of a few microkelvins. This is a well-known condition for gaseous condensates subjected to a magnetic field that is likely to occur at low temperature, as in the case of a quantum field of highly excited protons, at saturation, pulsed at threshold.

Schrodinger's non-linear liquidation of Gross-Pitaevskii [1.4] seems to me to be appropriateëe for estimating these interactions, the supernuidity ëbeing dëpendent on the frequency of the oscillations that result from the random interferences of the waves.

When a magnetic field modifies the interactions of the condensate by De Feschbach resonance, the atoms being highly correlated, low-energy collisions would then depend on the scattering length *a.* This parameter enabled us to determine the range of interactions between atoms. It turns out that when particles (proton, atoms) couple together, depending on this scattering length, the resulting waves interfere and diverge, within what limits?

When *a* is greater than 0, we have seen that the situation is repulsive at saturation, and when it becomes less than 0 it is attractive, to be cancelled out after the threshold for the triggering and coordination of cellular activity.

These hypothetical considerations on low biological energies oblige us to detail a primordial cellular organelle that we mentioned earlier, the mitochondrion, whose functions we will specify. The main one is cellular respiration, during which we observe the movement of electrons and protons in protein complexes located in their inner membrane. In plants, this process takes place under the direct effect of light during photon synthesis in the plant cell, and indirectly through a supply of nutrients for the animal cell, which is thus provided with potential energy during the anabolism and catabolism that contribute to its metabolism, by a sort of internalization of the action of photons, as free energy *(nuclear F)* for biochemical processes...

The mitochondrion (from the Greek *mitos* and *chondros)* is a cellular organelle that is thought to have originated as an endosymbiotic bacterium through the inclusion of an a-proteo-bacterium in a eukaryotic cell some two billion years ago. That of the order Rickettsiales (Anderson et A., 1998), which is fairly close to the ancestors of mitochondria, has acquired a specific genome that has been reduced during its evolution, and which is complementary to that of the cell nucleus. The mitochondrial genome encodes proteins assembled into highly complementary complexes destined for the internal membrane, and participates in its own transcription and in the translation of mRNAs.

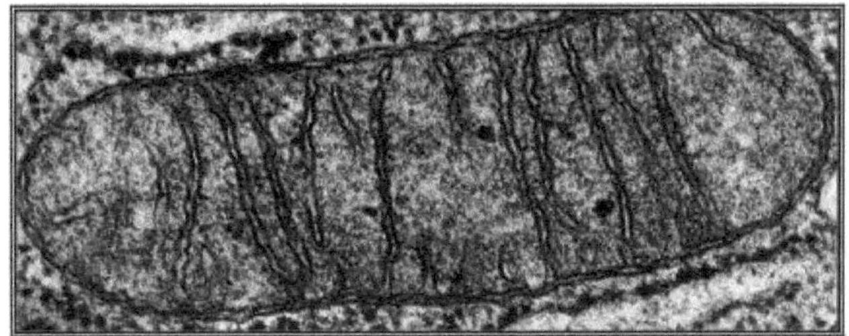

22. Mitochondria.

The mitochondrion, which looks like a filament and a granule, is mobile in the cytoplasm of the cell. This organelle, a few micrometres in size, is involved in the respiratory chain and the phosphorylating oxidation that produces ATP (Adenosine triphosphate[20]). This low biological energy process gives rise to a vital flow of electrons (e') and excited protons ($_{H+}$) that give up their free ёпе^e keeping the cell in ёnergёtic balance, reactive in the event of stress, a^rable during a pathological shock or extreme aggression. Stress has an oxidising tendency which produces free radicals causing cell death, particularly during apoptosis.

Depending on the cell type, the structure and number of mitochondria and their organisation in a network vary according to the ёtats ёnergёtiques and phases of the cell cycle.

The cell's ёnergёtic metabolism can be described as a complex chain of reactions, coupled and synchronised, if not coordinated with the cell's activity, to supply the organism with energy from its requirements for nutrients, тдёгёз, assimilates, dёgradёs, and excrёtёs. Energy comes from proteins, lipids and carbohydrates during catabolism, and is used to svntlK'tise substrates that contribute to the structural and functional anabolism of the cells and organism of each animal through various mёtabolic pathways.

These pathways are the Embden-Meyerhof pathway, which begins in the cytosol of the extra-mitochondrial cytoplasm and continues via the Krebs cycle[21] within the mitochondrial matrix, maintaining an ADP/ATP ratio which, depending on the organism's immediate needs, may be diverted directly to the pentose pathway. The mitochondria that bathe in the cell's cytosol are micro-organisms less than one to ten micrometres long and half a micrometre to one micrometre wide. A mitochondrion is made up of two superimposed membranes with specific permёabilitё and conductibil^ that confine two spaces and regulate exchanges with the cell environment.

---

20 In cold waters, Lithobathus sylvaticus frogs seek out waters rich in phosphorus, with a density greater than 0.4 mg/l.
21 Tricarboxylic acid cycle

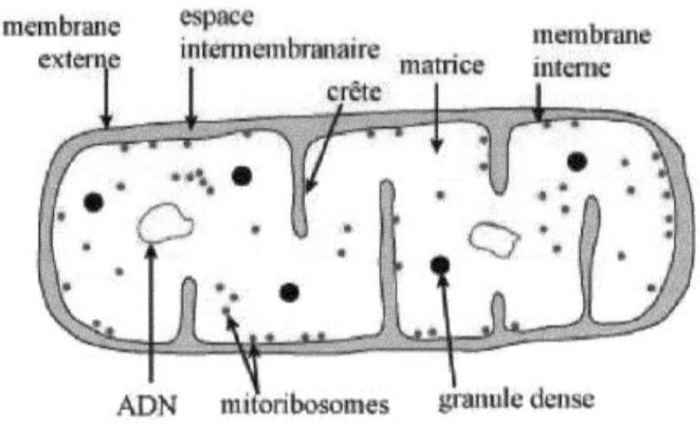

23. Diagram of a mitochondrion.

- At the centre of the mitochondria, the matrix, is the site of the citric acid cycle process, known as the Krebs cycle, which is a site of eight oxidation-reduction reactions carried out by enzymes. From glucose and pyruvate, acetyl coenzyme-A is completely oxidised, producing NADH[22] and FADH2.[23] Its weakly acidic environment is created by protons selectively entering and leaving the matrix via four of the five protein complexes in the inner membrane, which ëject them into the more acidic intermembrane space. From the matrix space and the inner membrane of the mitochondrion, ratsës are produced on the electron transfer chain, producing ROS (Reactive Oxygen Species) which play a role in apoptosis (cell death). To these ROS must be added electron and proton leakage, which is thought to be involved in various pathologies.

- The inner membrane that surrounds the matrix is formed by a series of crete-like folds that carry protein complexes, within which an assembly of complementary protein subunits have been formed that ensure the selective transport of ions and electrons, as well as the translocation of protons. I will go into more detail on the notion of translocation in the next chapter, as this process cannot be considered as strictly biochemical, but initially quantum, according to our model. The highly selective inner membrane is made up of 80% proteins and 20% lipids, and the intermembrane space is six to eight nanometres thick.

The movement of electrons and protons is vital for the maintenance of the mitochondrial respiratory chain, which depends on variations in the permeability and conductivity of the inner membrane that supports the five protein complexes. For each of them, we will detail the movement of electrons following the oxidation-reduction scheme, which produces a flow of electrons with apparently slow biological kinetics, and analyse the more discrete process of quantum and biochemical translocation of $H+$ protons, transferred from the matrix to the intermembrane space.

How do these highly excited $H+$ protons accumulate, initiating the threshold-pulsed quantum field with its high energy content, and how do they coordinate the cell's activity?

---

22 Nicotinamide adenine dinucleotide exists in the reduced form NADH, this redox couple is oscillating, the energy of these oxidations produced in the internal membrane gives an electrochemical gradient inducing a proton gradient.

23 Flavin adenine dinucleotide forms a redox couple with FADH2. FAD participates in the formation of ATP by pumping six protons into the inter-membrane space.

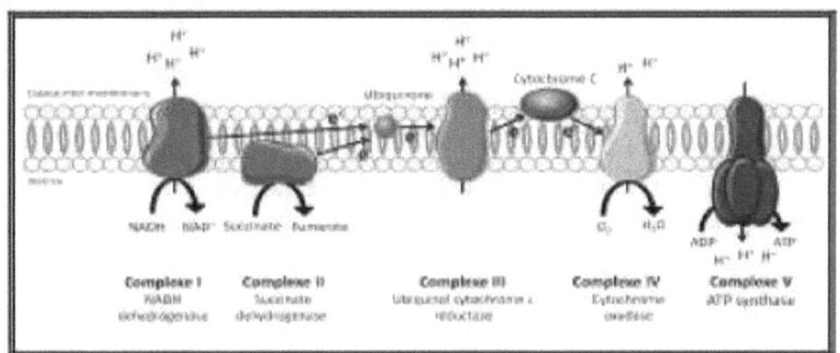

24. Inner membrane and protein complexes, electron flow and proton translocation.

. **Complex I**, NADH-ubiquinone oxidoreductase, has a structure of forty-four subunits, initiates electron transport and translocates four protons into subunits positioned in the inner membrane. Its L-shape is extended into the matrix by a hydrophilic arm of eight oxidation-reduction subunits and nine iron-sulphur oscillators[24] oscillators linked to the Flavin Mono-Nucleotide.

$$NADH + Q_{10} + 5H^+_{matriciels} \rightarrow NAD^+ + Q_{10}H_2 + 4H^+_{intermembranaires}$$

After this dehydrogenase of **NADH**, the redox balance uses the reversible oscillator $Fe^{3+} \rightarrow Fe^{2+} \rightarrow Fe^{3+}$, the iron atom of the protéine cytochrome, to finally give **NAD'** and $_{H2}$ **O.**

Décrivons précisëment le processus du complexe **I** qui transfère deux ëкйго^, c'est la réduction de l'ubiquinone en ubiquinol par un gain dklectrons regu du NADH qui modifie les changements conformationnels du bras membranaire ou se déroule la translocation des protons. A process, initially quantum in application of Schrodinger's equation, subsequently biochemical and thermo-hydro-dynamic, by transferring four protons to the inter-membrane space. This accumulation of **H+** induces a strengthening of the **pH** in the inter-membrane space of the mitochondrion, which would change from low to high during the long mëtabolic phase preceding the short activik phase of the cell according to our research livpotliese dëveloppëe in chapter I, namely :

"The threshold-pulsed quantum field would spontaneously reset the conformity of the internal and external protein-membrane complexes, and coordinate cellular activity in its triple aspect - quantum, random and discrete - which would be faster than the biochemical reactions to thermal and hydric effects. The dynamics of these fluctuations-dissipations would mask the quantum process, which would be spread out over time, through its action, due to the interference of the resulting divergent waves, on the dynamics of the protein complexes, and more generally on cellular and organismal activity".

In this case, the cëlëritë of trans-localised protons at the subunits of the protein complex I (in Homo sapiens, ND5, ND4, ND2) linked together by an HL amphiphatic helix, would produce an energy transfer, according to these two complementary hypotheses.

1- Quantum hypothesis.

If we consider the energy released during oxidation-reduction as the potential energy of the flow of electrons that ensures non-localised bonding with the excited protons **(H+)** located in

---

24 The role and excess of iron are discussed in 'L'Euprocte des Pyrenees' associated with this essay. Iron and sulphur are cited by J.D. Bernal, J.B.S Haldane, N.W. Pirie and J.W.S. Pringle in '*A Discussion of the Origin of Life*' (1955).

the matrix. These excited protons are in search of electrons. As a result, under the influence of this negative potential known as the proton-motive force, a proton with a non-homogeneous positive electric charge is directed towards a protein sub-unit with an available negative charge, in a sort of electronegative proton tunnel. This potential difference *(A y)* would tend to accelerate the proton *(H+)*, enough to cross the potential barrier of the protein subunit, by tunnel effect, to be finally ejected into inter-membrane space in an over-excited state of corpuscle *(H+*)* and wave, according to Schrodinger's equation, two aspects under which it must be considered for our demonstration.

2- *Biochemical hypothesis.*

When quinones are reduced, the energy released causes a conformational change in the subunits which, like a biological motor, modify their inclination by reorienting the active residues, making them capable of trans-localising four protons on carrier sites.

The translocation subunits and the composite helix are simultaneously subjected to the electron field and the energy released by the proton in a weak biochemical bond that would not actually bind the proton, if considered in its quantum wave state, to the translocation sites thus formed.

To put it simply, this assembly would act like a gear into which the proton, both corpuscle and wave, would be inserted, driven by a ratchet which, by its momenUine arrangement, *would "open"* the proton tunnel, which would easily and very quickly cross the potential barrier by virtue of its transparency. It would reach the inter-membrane space where the excited protons (*H*) would accumulate, and the resulting acidity, which would vary from weak to strong *(A pH),* would create a pulsed field of highly excited protons and photons. These highly excited protons and photons (*H+*) would interact with each other and with the atoms in the mitochondrial environment, through a range of resonance frequencies, up to a threshold of rearrangement. This process would trigger the activity and coordinate the cellular exchanges that are essential for the immediate supply of nutrients to ensure that the oxidation-reduction potential is restored after the instantaneous rearrangement of membranes and protein complexes.

This quantum field and its resulting divergent wave problems would thus act as a coordinating unit for cellular activity, in the latter case in application of Schrodinger's non-linear equation.

. *Complex II*, Succinate-ubiquinone-oxidoreductase contains four protein subunits in the Krebs cycle and catalyses the oxidation of succinate to fumarate. Electrons are transferred from succinate to ubiquinone without proton pumping, as the energy is probably insufficient for proton tunneling due to the inappropriate configuration of the subunits. The comparative and evolutionary study of this complex in relation to the others seems to me to be an interesting way of assessing how the proton translocation modalities were formed during the revolution of the animal cell, with the exception of this complex.

. *Complex III,* Ubiquinol-cytohrome c oxidoreductase (BclComplex) is positioned in the inner membrane, receives the electrons from ubiquinone, this oxidation causes the translocation of four protons and the ubiquinol electrons transposed by cytochrome c via the oscillator (Fe^{3+} / Fe^{2+}) are directed towards complex IV.

Inhibition of ubiquinol oxidation neutralises the proton gradient and ATP synthesis, while causing the consequent production of[25] superoxide radicals. We can consider this to be a

---

25 Several pathways are being explored by researchers for the production of free radicals, and their incidental formation could take place at the level of pyruvate dehydrogenase, before electrons enter the transport system of

critical site for vital coherence, generating chain reactions with pathological consequences.

. Complex IV, cytochrome c oxidoreductase, at the end of the redox chain, uses the electrons from cytochrome c (bimetallic site) to reduce oxygen to molecules of water, while trans-localising four protons.

$$4\ CytC\ Fe^{2+} + O_2 + 8\ H^+ \rightarrow 4\ CytC\ Fe^{3+} + 2\ H_2O + 4\ H^+$$

These four complexes ensure the potential of the inner membrane $(\Delta\Psi)$ and the differential proton gradient, the field of which is localised from the matrix to the inter-membrane space (ΔpH) . The protons are used by ATP synthase, which catalyses the formation of ATP by phosphorylation of ADP in complex V, a veritable low-energy biological motor.

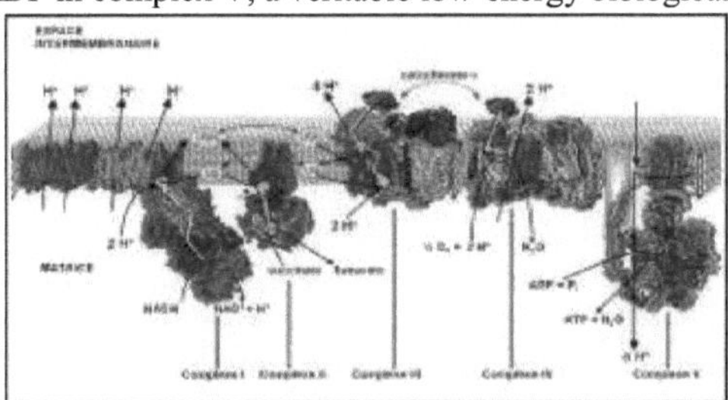

25. Dëtails of the subunits of protein complexes
. The ATP synthase **complex** is the seat of phosphorylating oxidation, its twenty or so sub-units capturing the energy of the inter-membrane proton gradient (ΔpH) This proton driving force and the updating of the ADP/ATP ratio condition its direction of proton translocation, directed either towards the interior of the matrix or expelled towards the inter-membrane space in coordination and synchronisation with the other complexes. Its eight specific subunits $(\alpha,\ \beta,\ \gamma,\ \delta,\ \varepsilon,\ a,\ b,\ c)$ make it a veritable biological low-energy motor. The **F-type** ATPase is made up of two parts.

. Fo, positioned in the membrane, translocates H+ protons using its trans-membrane sub-units "c", Y and s, which are negatively charged. Initially, a quantum process is produced by the action of the proton force driving the flow of electrons which acts on the translocation of protons in Fo, towards Fi, whose *"spinning"* produces the phosphorylation of a molecule of ADP into ATP in the three catalytic B subunits.

. Fi is in contact with the matrix, its active globular protein head is composed of three catalytic sites:
. A fixed part similar to the stator of a motor
. A moving part dësignëe rotor
. the inter-membrane space
- The outer membrane of the mitochondrion, dëlimits the inter-membrane space and extra-mitochondrial ëchanges in contact with the cytoplasm of the cell. This membrane is a sëlective membrane composed of a lipid bilayer whose porins are ion channels for the passage of metabolites (fatty acids, nucleotides) and ions by diffusion, or for active transport by a translocase. It should be noted that this membrane remains impermeable to H+ ions, which

---

protein complexes.

accumulate in the intermembrane space, creating the $H+*$ threshold pulse proton field. This field would change from a weak to a strong force during the metabolic phase, and would be expressed at saturation by an effective wavelength on the differential rearrangement of atoms and molecules, triggering and coordinating the vital activity of cellular exchanges, but also inducing electroporation of the external membrane sensitive to different frequencies.

In addition, from the random relativistic point of view, we need to consider the multi-causal hypothesis that an excess of protons and their virtual photons could modify the effective wavelength of the threshold pulse signal, to such an extent that the interference of the resulting waves could reach a state of critical divergence likely to insidiously disrupt enzymatic and cellular functioning, for example by disturbing ferromagnetic oscillators or alkaline ions.

The simultaneous presence of these divergent waves of the $H+*$ proton field and free radicals, which are electrically neutral but have paramagnetic susceptibility, also raises questions. Could this configuration be at the origin of deleterious chain reactions?

Understanding this potential disorder is important, because protein complexes are involved in the fission and fusion of mitochondria, and provide the link with the cell's cytoskeleton and the proteins involved in lipid synthesis and transport. In comparison, the inner membrane is even more selective in terms of the permeability of ions, which have to pass through biochemical transporters rather than pores. These complex proteins are embedded in a multitude of hollows, the number of which indicates the level of cellular activity. The structure and number of the mitochondrial network varies according to cell type and tissue, giving an idea of the different energy states and phases of cellular activity. Apoptosis contributes to the situation we call out of stable equilibrium for the maintenance of the organism's homeostasis, which is in a statistical state that affects it from relative stability, under genetic and epigenetic control, to random and stabilising variability, depending on the energy from the mitochondrial protein complexes. This instability normally manifests itself during embryogenesis, metamorphosis and incidental organ regeneration. It is this structuring and/or destructuring aspect of a fundamentally random nature that is thought to contribute to the vital low-energy biological coherence generated by the movement of electrons and protons in the mitochondrial inner membrane during cellular respiration. We have described the inner membrane that supports five protein complexes, each of which is a true composite unit of integration level.

After the Krebs cycle reduction or beta oxidation in the matrix, four complexes host a succession of oxidation-reductions during which NADH and FADH2 are re-oxidised. Through these successive oxidation-reductions, the transport of electrons is correlated to the translocation of protons during ATP synthase in complex *V* and in the trans-membrane protëine complexes *I, III, IV*, to the exclusion of complex *II* which does not transport any proton.

This translocation of protons from the matrix to the inter-membrane space via these protëine complexes generates the membrane potential $(\Delta \Psi)$ and gives rise to a difference in ΔpH which tends to increase in the inter-membrane space from a weak to a strong acid force. The nominal proton-motive force *Δ p is given by* the following formula.

$$\Delta p = \Delta \Psi - (2,3RT/F)* \Delta pH \qquad [2.14]$$

R corresponds to the perfect gas constant 8.314 J.mol^{-1} .K^{-1} T is the temperature in Kelvin.

F is Faraday's constant, i.e. 96 485 C.mol^{-1}

In terms of energy, a transmembrane **pH** difference of one unit is equivalent to a potential difference of 60 millivolts. In the mitochondrion, this value is between 150 and 200 millivolts for a **pH** difference of 0.5 at the most. The difference in the electrochemical potential of $H+$ ions is known as the proton gradient, and is considered to be the thermodynamic force behind ATP synthesis.

This process corresponds to the chemo-osmotic theory described by the English chemist Peter Dennis Mitchell (1920-1992), according to which the coupling of redox reactions and phosphorylation induces proton translocation concomitant with electron transport. Acidification of the inter-membrane space *(pH=6.2)* relative to the matrix *(pH=7.6)* causes phosphorylation of ADP to ATP, at a potential difference of 120 millivolts. This theory will be completed by that of the biochemist Paul Delos Boyer (1918-2018) and the chemist John Ernest Walker, associated with the 1997 Nobel Prize winner, the physicist Jens Christian Skou (1918-2018). Their concept of rearrangement and ion transport was related to the framework of Faustino Cordon's theory of integration level units. His daughter Teresa hypothesised the role of the proton field in the pulsation and coordination of cellular activity after the long metabolic phase, which precedes a short phase of cellular activity, by causing the inflow of nutrients and the outflow of waste products.[26]. According to Teresa Cordon :

For cellular action to be the result of somatic protein activity,
sets of somatic proteins must establish a more rapid and integrative coordination than those that establish contiguous proteins between them".
a more rapid and integrative coordination than that established between contiguous proteins".

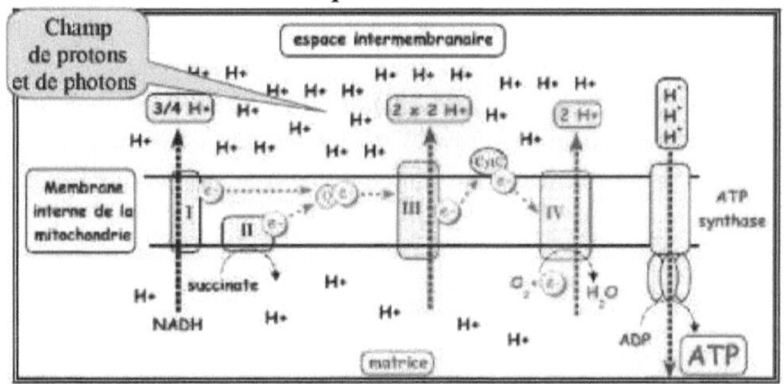

26. Field of protons and photons at saturation in 1
intermembrane space.

In this case study, according to my naturalistic research hypotheses, the proton potential accumulated in the intermembrane space would change from a weak to a strong acidic force, like a pulse wave at saturation, which at threshold would instantaneously reset the atoms and molecules of the mitochondria and the cell by modifying its permeability and conductivity in order to *'feed'* the long metabolic phase.

This hypothesis requires us to consider the quantum aspects mentioned above, in order to demonstrate a higher degree of ceratogenicity than the biochemical and thermohydrodynamic

---

26 This process takes place in the skin cells of amphibians
in direct contact with the external aquatic or aerial environment.

effects that take place under the control of ATP synthase as a function of the ADP/ATP ratio, which depends on the immediate energetic needs of the cell and the organism to feed itself in relation to the selective constraints of its environment.

Without going into the mëandres of quantum physics we propose a test model on the prësumë role of the proton-threshold pulsed quantum field in ensuring the transition between the nanoscopic, the microscopic and the macroscopic. These different perceptions of integration level units hint at singularities of a nature mainly from the statistical domain of quantum physics to the underlying, discrete and random effects, indistinguishable from the processes of thermo-hydro-dynamic fluctuations and dissipations, during the discernible activity of the cell

.

The functions of the mitochondria are essential for phosphorylating oxidations and are involved in the metabolism of sugars as well as other important metabolic functions, such as :

- Iron metabolism, a fundamental oscillator, omnipresent in protein complexes.
- Lipids and amino acids.
- Calcium signalling.
- The synthesis of steroid hormones.
- cell death, apoptosis, involved in cell homeostasis, embryogenesis, metamorphosis and organ regeneration.

We also need to take into account the leakage of electrons and protons due to the properties of the inner membrane.

I would like to quote two forward-looking hypotheses[27] on this subject:

- The primary cause of leakage is thought to be induced by thyroid hormones, linked to an increase in the surface area of the inner membrane. The sensitivity of iodine to light is shown below. Hydrogen iodide **HI has** an absorption spectrum between 207 and 360 nanometres wavelength, and its dissociation produces free radicals *(°)* whose quantum yield corresponds to the number of molecules decomposed in relation to the number of photons observable at energies greater than the binding energy of **HI.** Sunlight is captured by plants during photosynthesis, while the photons that pass through the foodingers have become difficult to discern in animals.

$$HI + Hv \rightarrow H^{\circ} + I^{\circ}$$
$$H^{\circ} + HI \rightarrow I^{\circ} + H_2$$
$$I^{\circ} + I^{\circ} + M \rightarrow H_2 + M \qquad [2.15]$$

In the presence of water that can be dissociated into H_2 and **O**, the following reaction also produces a hydrogen radical.

$$I + H_2 \rightarrow HI + H^{\circ} \qquad [2.16]$$

With these possibilities in mind, we now need to look at the branched chain reaction of iodine, which tends to accelerate biochemical decompositions in very small quantities, *while at the same time regenerating.* Its concentration affects the speed of reactions and the very low activation energy, which induce explosive chain reactions if they are amplified, and are therefore normally contained to low biological ë energies.

. At amorgage we have, $I_2 \rightarrow 2I^{\circ}$

. After the thermolysis chain reaction with another ëlëment, when the divergent limit process

---

27 In the fourth volume of the naturalist suite, Indri indri, voyage aux origines de l'Humanite, I discuss the role of iodine in the hominisation process.

breaks down, we obtain $2I^- \rightarrow I_2$, which returns to its original chemical form.

- The second cause of electron and proton leakage comes from UCP proteins capable of transporting protons and involved in thermogenesis. These coupling proteins, such as UCP1, are present in large quantities in brown adipose tissue.[28] full of mitchondria.

At the end of this chapter on the mitochondrion and its vital functions due to the transport of electrons and the translocation of protons, it seems important to me to know how these metabolic functions react to a quantum field saturated with protons and photons, pulsed at threshold, in order to coordinate cellular activity, taking into account the normal and abnormal effects expressed at different levels of resolution, proteins, cells, animals...

---

28 I have dealt with brown adipose tissue in Cevennes beavers, which are subject to severe thermal and hydric stresses when rivers dry up seasonally.

Chapter 3

Coordination of cell activity

The protons expelled from the matrix by protein complexes *I, III and IV* of the inner membrane into the intermembrane space of the mitochondrion are protons that are *'accelerated'* by the flow of electrons generated by the redox reactions taking place in the five protein complexes. These protons give up some of their energy when they are translocated to rearranged carrier sites. By accumulating in the inter-membrane space during the long metabolic phase, this *H+* field increases the acidity of this space for a period that corresponds to the differential resonance of this field of *H+** protons interacting with the atoms of the mitochondria and the cell according to their magnetic susceptibility, and paramagnetic susceptibility for free radicals. When the composite proton-photon-atom field reaches saturation, it is thought to be the cause of the threshold triggering and coordination of the cell's activity phase, determined by the imperative need to feed, a vital need that leads to this action.

At threshold, the resulting pulse field would instantaneously modify the conformation of the atoms of the inner membrane and protein complexes according to their magnetic susceptibilities by acting on the orientation of the spins, which would modify their energy levels into diifereneated hyper-fine structures. This spontaneous rearrangement would transform the permeability and conductivity of the membranes of the mitochondrion and the cell, whose instantaneous field, with its high celerity of fermions and bosons, would effectively ensure the unity of its coordination. We have retained this hypothesis for our test model.

To explore the transition of the energy levels of the hydrogen atom to form *H+*, I based myself on the studies cited earlier by Glass-Maujean (1974) who demonstrated hyperfine structures by dissociating the *H2* molecule using a beam of slow electrons which collide with *H2* at low pressure in a closed medium. From his local study of the Zeeman diagram, I retained the n=3 and n=4 excited levels of the hydrogen atom, in relation to the values of a magnetic field that shows two Zeeman sublevels of the same energy crossing each other. By applying a perturbation such as that of the pulsed field of *H+* protons, which tends towards *H+** saturation at threshold, these two sublevels would be repelled and would intersect, introducing transitions. After his description of the fine structure of the hydrogen atom using Dirac's equation, Pauli Darwin's equation seemed more advantageous for this type of research.

-   The *Φ* wave function has the advantage of having only two components.
-   *Eobs* , corresponds to the observable binding energy.
-   *Eobs<!>* includes corrective terms, including the Darwin term.

$$((p^2/2m + eV - p^4/8m^3c^2 + e\hbar\sigma.(VV^\wedge p)/4m^2c^2 + e\hbar^2/8m^2c^2 \blacktriangledown^2 V))\ \Phi)$$

[3.17]

It is the variation in this binding energy that interests us in the context of the translocation of excited *H+* protons through protein complexes, which accumulate, trapped in the mitochondrial inter-membrane space.

-   The first two terms, *p2/2m + eV,* characterise the non-relativistic Hamiltonian.
-   The third term, *p /8m*[43] c2, expresses the relativistic correction.
-   The fourth term, *ehe.(VV^p)/4mrc* , represents the spin-orbit interaction.

- Darwin's last term, $e\hbar^2/8m^2c^2\,\nabla^2 V$, would act exclusively on the **s** states in question, specific to the "***non-localisation of the Dirac electron***", which in this case would be the mean value of the electron flow. It would include a part oscillating at relativistic frequencies = $2mc^2$ and amplitude = h/mc, by the Zitterbewegung effect, i.e. at the relativistic limit, the interaction between the electrons and the proton potential would no longer be local. This non-localisation would find its application in the oscillating variation of the binding energy between the flow of redox electrons and the trans-localised $H+$ protons, as a proton-motive force acting during the long metabolic phase:

. Firstly, on the extraction of the proton matrix in exces by a quantum translocation of the Schrodinger wave type, towards and in protein complexes, by tunnel effect.

. Secondly, by a relatively more localised relationship at biochemical acceptor sites made up of the subunits of protein complexes, periodically rearranged to trans-localise these protons from each complex to the intermembrane space.

On the excited hydrogen atom, precision measurements gave a lamb-shift frequency of 5.88 +/- 0.65 Mhz at level 3 and 2.2 +/- 1.0 Mhz at level 4. This experiment highlighted the formation speeds of the different levels during the dissociation of the H2 molecule, estimated at between 17 and 25 eV. At the excitation threshold, the speed of the atoms produced was 8.0 +/- 0.3 km/s for a pre-dissociation energy of 17.23 +/- 0.04 eV.

In other words, two pre-dissociative processes that serve as a reference for our field of application.

- One, for an energy of between 17.2 and 18 eV, where slow atoms are formed, at around 9 km/s. This is the assumption we use for the quantum and biochemical translocation of $H+$ protons in protein complexes *I, III, IV* and *V*, in application of Schrodinger's equation.

- The other, for an energy of 25 to 33 eV, gives fast atoms of 30 to 40 km/s. Hypothesis of the initial velocity dedicated to the field of pulsed protons at threshold, becoming highly excited H+* until they reach relativistic velocities according to *[3.17]*, once they reach saturation, which would correspond to the application of Schrodinger's non-linear equation.

Isotopic effects must be considered, particularly on the more discrete and random sëlective rearrangements at atomic levels and consëctively at the molecular levels of protëine complexes, membranes and components of the cellular milieu :

- On the one hand, on the capacity for quantum translocation in the dissociated form *H* under the impulse of the flow of electrons, at the speed of 9 km/s, which cëde its free energy by establishing the biochemical translocation bonds of each proton with classical thermos-hydro-dynamic consequences by fluctuations-dissipations on the regulation of temperature, remarkable more particularly at very low temperature.

- On the other hand, the formation of the threshold proton pulse field with an initial speed of 9 km/s at low *pH* increases to an average of 40 km/s at high *pH*, reaching relativistic speeds, when the threshold proton and photon pulse field, which at saturation reaches a resonance maximum with the atoms in the mitochondria and the cell medium, rearranges itself by the Zeeman effect.

Initially, the process would be exclusively quantum and indistinguishable due to the celestial nature of the pulsed proton field, depending on its frequency and relativistic wavelength, with the production of discreet interferences and divergences. Let's try to appreciate the consequences between the nanoscopic, the domain of cold solutions, the microscopic and the macroscopic, that of hot solutions, of the units of integration level considered, proteins, cells, animals.

Using the model approach of the preceding chapters, we have given an account of this process, which effectively brought out the ëvidence.

-Strictly quantum cold singularities, which are expressed by wave couplings at the level of protein complexes and the inner membrane of the mitochondrion.

- From fluctuations-dissipations of indiscernible quantum origins to the biochemical extensions of hot solutions discernible in the classical thermo-hydro-dynamic domain of the cell.

To recapitulate, to meet the cell's imperative need to supply itself with energy nutrients by reinitialising its vital oxidation-reduction process associated with the translocation of protons.

The biological cell's respiration, which takes place in the protein complexes of the inner membrane of its mitochondria, is thought to generate a unitary $H+*$ field. This field, formed during the long metabolic phase, once it reaches saturation, or the threshold, triggers and coordinates the cell's activity by provoking rearrangements favourable to exchanges between the cell, its immediate environment and its surroundings, in order to feed itself and maintain the flow of electrons in conjunction with that of protons. These protein complexes bring energy to the cells, whose genetic and epigenetic systems regulate their structuring and/or destructive capacities, evolving at each unit of integration levels, emerging and different from the previous one, while maintaining the animal's vital coherence.

In the course of a species' contingent devolution, its adaptability depends on fundamentally random variations, initially quantum in the space-time of their resolution at the level of the animal's proteins and cells, which are expressed by a stabilising variability of genetic mutations and epigenetic interactions in their complex, effectively Darwinian, relationship to the randomness of natural and anthropic selective constraints, in space and time.

There are two test modëlisations a opërer to discern these complex processes:

- During the long mëtabolic phase, consideredërëe as a systëme fermë, the first modëlisation corresponds to the transport of ëκ^го^ together with the translocation of protons into inner membrane prokine complexes. We propose a ëlëmentary sclwina that combines aspects, quantum, biochemical and thermos-hydro-dynamics.

- During the short activik phase, as far as the field of saturated protons is concerned, pu^ has a threshold, which would correspond not only to its prësumëe capacitance for coordinating cell activik, of the open system, but also to its normal structuring and/or dëstructuring capacitances, and incidental ones which remain to be explored, in particular on the expression of gënes.

"As a function of the degree of divergence of the derived waves which would be expressed at long range, independently in the
dimensions of space-time".

Let us begin with the simplified test modëlisation, to modëlise the pи^ë field that would build up in the intermembrane space of the mitochondrion during the long mëtabolic phase that precedesëcë its shorter activ^ phase. As the proton field builds up, from the end of translocation into the intermembrane space of $H+$ at 9 km/s, after interaction of the $H+*$ protons it would increase to 40 km/s to reach a brief relativistic velocity at saturation of protons and photons, interacting with the atoms, up to the threshold of deëclenchement which would cause coordination of the activ^ of the cell. For the modëlisation, we consider this Poisson law process as a function of their initial relativistic velocity of 40 km/s, in a closed environment such as a wide-frequency black body.

or, $[40H^+]^3/(e^{[H++]} - 1)$ $\qquad\qquad$ [3.18]

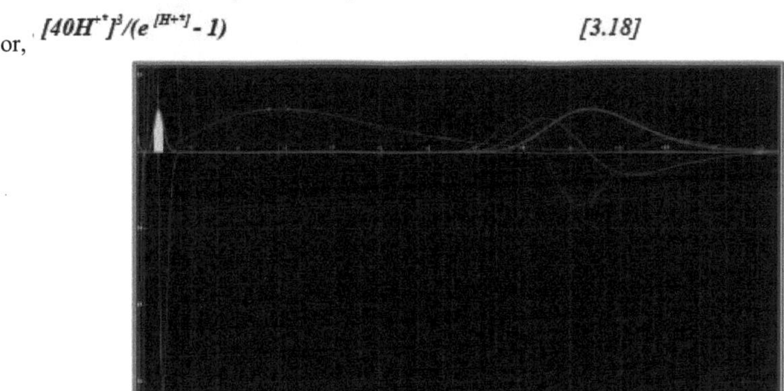

27- Modëlisation of the distribution of the pu^ field of protons and photons at saturation, in the intermembrane space of the mitochondrion.

-   The red curve represents the probability distribution^ of the proton field at an initial speed of 40 km/s, which reaches a maximum at a relativistic speed of saturation of protons and photons (green arrow).

-   The blue curve corresponds to the forging by its wavelet transformëe, to simulate its spectrum of energy levels, relativistic at saturation, composedë of fermions and bosons whose diffusion length *a* is analysed, starting from the first zëro of the axes.

. On the left, the positive peak, should be considered in the strictly quantum domain with its ^gative dërivëe, this singular potential well (shaded area on the left) for the "*cold solution*" would correspond to a pseudo-condensate, dëlimitë by the points of inflection and the maximum of the positive peak whose intëgration gives a dens^ of probabil^ close to 10, which would be an indiscernible fleeting effect.

-   On the right, the field dëcroit and is manifested by a normal distribution, in the thermodynamic domain, with ëquilibration of tempëratures, a "*hot solutions*", representedësentëes by its first and second dërivëes.

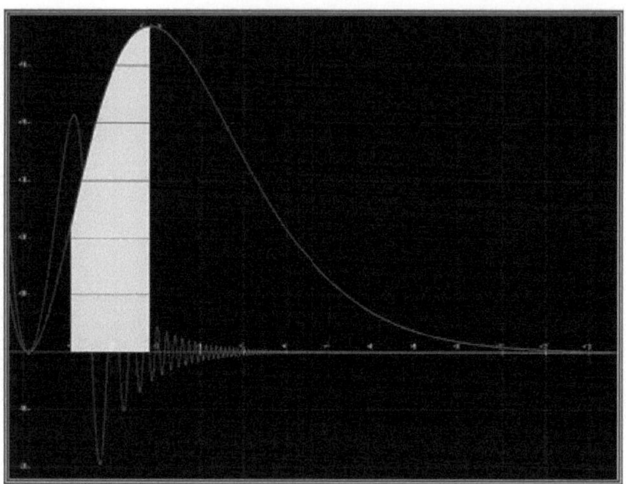

28- Simulation of the proton pu^e wave, at saturation.

- On the red curve, the пПёдгё space of the field of protons and photons at 40km/s has reached saturation, at its relativistic maximum, the inlegration gives a dens^ of probabil^ of 83, for 10 ёvaluё compatible with a pseudo-condensate.
- the blue curve simulates the pu^e wave of protons and photons at saturation, i.e. :

$$83 sin\ [H^-])^3 /(e^{[H+]}-1) \qquad\qquad [3.19]$$

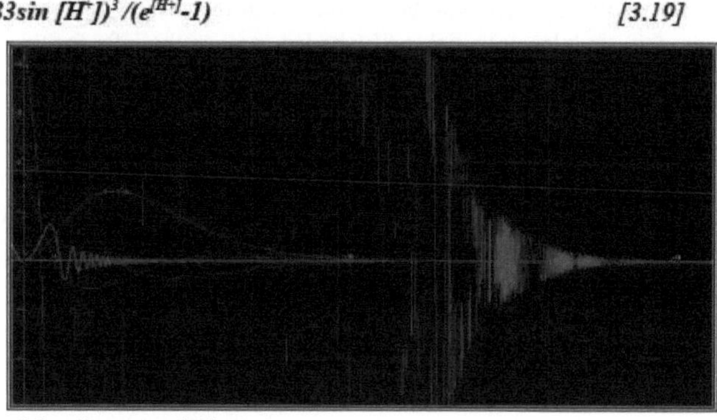

23- Simulation of the divergent dёrivёe wave.

- The blue curve represents the pulsed wave. With saturation of protons and photons, the first derivative shows local maxima (vertical blue lines on the left) and the second derivative cancels out into a problematic divergent wave (on the right).

At the threshold, how does this wave interact with the atoms of the inner membrane and its protein complexes?

For an average of the fluctuations of the membrane $1/\sqrt{N_{atomes}}$ and the protein complexes $1/(ln\sqrt{N_{atomes}})^2$, we have several solutions following the coupling of the resulting waves.

For internal and external membranes :

83 sin $(x_{protons})^3/(e^x-1))*(1/\sqrt{x_{atomes}})$ *[3.20]*

for x protons and atoms in the inner membrane.

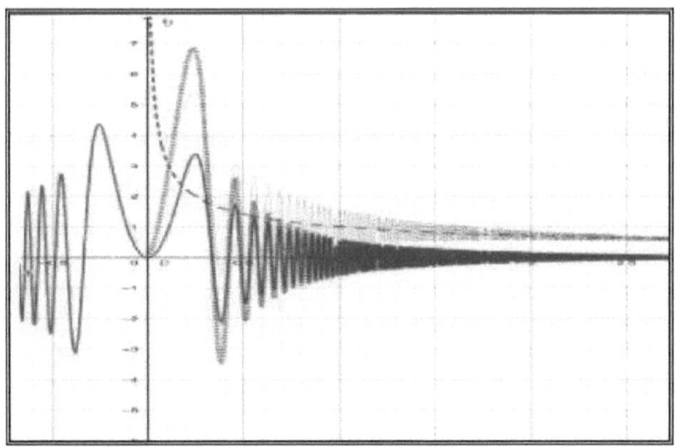

29. Simulation of 1 pu^e wave of protons and photons at saturation
(in red) which interferes at threshold

with the atoms of the inner membrane (dashed green).

Two solutions are envisagëes:

-For the ***multiplicative cold solution*** (in violet) by aligning with the initial wave we see a divergent solution in excess (positive violet peak) which cancels out, this potential well would be the indicator of a pseudo-condensate mentioned^ in the prëcëdent diagram.

-The dëcalage of the ***additive hot solution*** (in green dots) would give a divergent solution stabilisëe by absorption of the wave towards regimes of thermo-hydrodynamic fluctuations and dissipations.

For inner membrane protein complexes :

83sin $(x_{protons})^3/(e^x-1))*((1/ln(\sqrt{x_{atomes}})^2)$

For x protons and atoms of protëine complexes. *[3.21]*

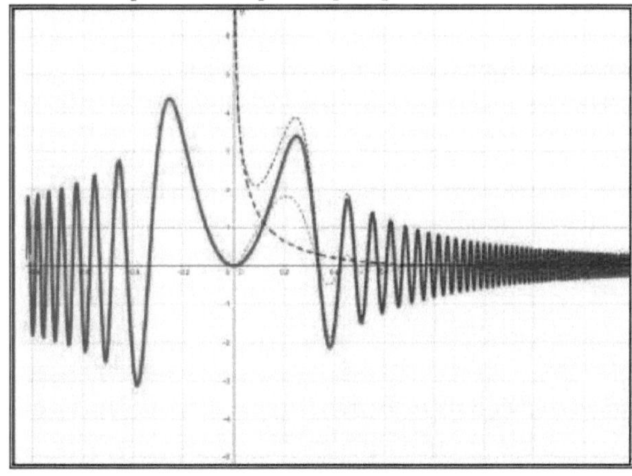

30. Simulation of the proton wave pulsed at threshold (in red),
interacting with protein complex atoms (
blue dashed line
).

Let's look at the two solutions:

- ***The multiplicative cold solution***, the violet peak that cancels out, would indicate a pseudo-condensate.

- ***The additive hot solution*** adopts a thermo-hydro dynamic regime.

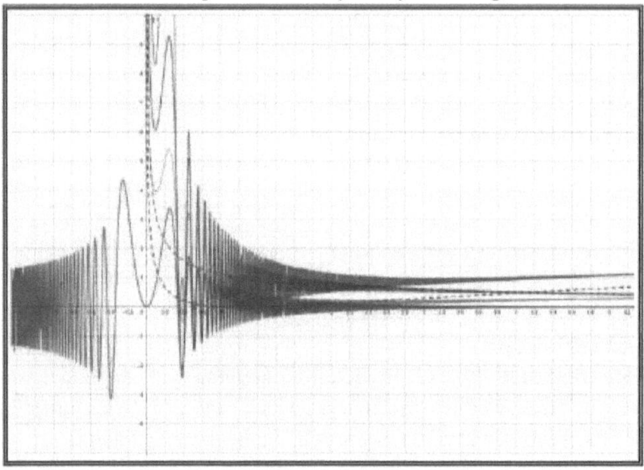

31. Simulation of the pulsed wave of protons and photons (red),
at saturation, in threshold interaction with atoms

of the inner membrane and its protein complexes.

If we observe all the couplings, the proton and photon field at saturation shows two peaks (negative and positive) of condensation. The peak on the right is absorbed by the membrane and the protein complexes in a superposition of initially indistinguishable cold quantum solutions, then discernible hot solutions that are superimposed in a thermo-hydrous-dynamic regime of fluctuations-dissipations that mask the cold solution.

After this visual approach, we need to introduce the formalism of quantum physics, and we move on to the test model using the nonlinear liquidation of Gross and Pitaevskii.

for

$$\Psi = sin(83x^3)/(e^x-1). \qquad\qquad [3.22]$$

$$-id_t\Psi + \Delta\Psi + \Psi(1- |\Psi|^2) \qquad\qquad [3.23]$$

Ψ *.wave.*

$|\Psi|^2$ *density of the wave.*

$\Delta\Psi$ *average variation of the wave in the trap.*

- for the first term, - $-idérivée_t ((sin(83x^3)/(e^x-1)).$ [3.24]

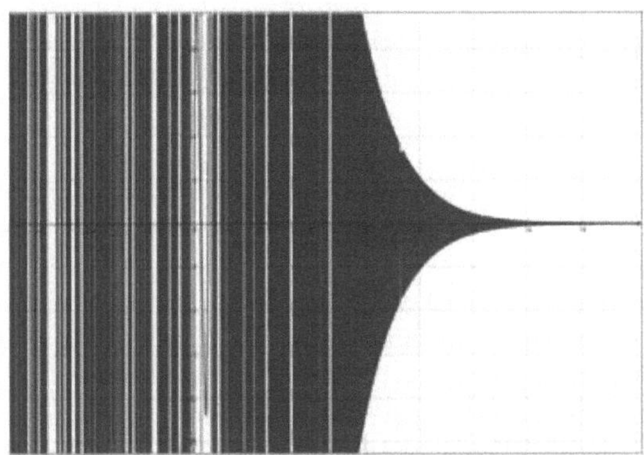

32. The decisive wave is highly divergent, with
phases of intermittency in white, particularly around the zero point of the axes, a
zone of cold and hot
solutions detailed in the
previous graphs.

For the second term, we use the average fluctuations, which trap the protons of the internal membrane $(1/\sqrt{N})$ or the average fluctuations of the *In* protëine complexes $(1/\sqrt{N})^2$ which constitute the external liquidation constraints.

$((sin(83x^3)/(e^x-1))((1/\sqrt{x})+ ln\ (1/\sqrt{x})^2)$ for x protons and atoms.

- Finally, the third term expresses the non-linear trend

$$((sin(83x^3)/(e^x-1))((1- (sin(83x^3))/e^x-1)^2) \qquad [3.26]$$

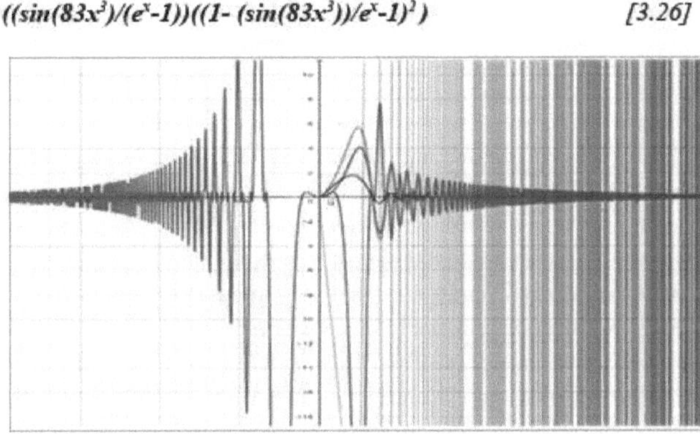

33. Simulation using
Gross and Pitaevskii nonlinear liquidation.

The coupling of these three terms gives the probable solutions of the pulsed wave (violet), the proton and photon field at saturation, and at threshold, during interactions with the atoms of the inner membrane and its protein complexes.

The consëquences of its divergence and the potential wells (green) as well as those of the coupling modes (blue, red, orange) remain to be defined.

Note, the Mexican hats (very sharp green peaks) that spread out around zëro, which we have analysedë prëcëdently as ëbeing indicative of an indistinguishable fleeting pseudo-condensate, momentarily has ëlevëe entropy and low ëenergy.

This state would be a source of vortexes, i.e. whirlpools at the origin of micro-states, like potential seeds of growth and development, or regeneration.

To summarise this chapter, during the cell's metabolic phase, protons leave the matrix via complexes I, III and IV and return from the intermembrane space to the matrix in complex V, which regulates the ADP/ATP ratio according to the animal's energy requirements, as it is forced to feed.

- The flow of electrons as potential energy acts on the protons by causing a tunnel effect, this proton driving force extracts the protons from the matrix which are trans-localised via optimised acceptor sites during the cycles of the long metabolic phase.

- The binding energies of the electron flow polarise the proton, which would justify its translocation to the site of the optimised carrier atom. We consider this complex of non-localised binding energies as an oscillator sensitive to perturbations according to the Schrodinger equation.

For low tempëratures between 0.1°C and 1.5°C, relative relocation of the proton from its acçeptor would give hyperfine structures, a wide range of energies that characterises the state of the oscillator, according to Darwin's term *[3.17]*. Depending on its oscillating energy level, the proton accelerated by the electron flow and slowed down by the acceptor atoms would continue its trajectory, in this tunnel, towards another acceptor before ending up in the intermenbranary space, where they form a field of highly excited protons and photons, at saturation.

According to our model of the non-linear equation of Gross and Pitaevskii, the coordination of the cell's activity is operated by superimposed fields, created from the strictly quantum entanglements of protons and photons, where the notion of space and time merge.

- A first state would ensure the formation of the field which would go from weak to strong, in a combination of protons and photons at saturation, their interactions would reach relativistic speeds.

- the second state would correspond to the instantaneous coordination of the cell's activity, by the action of this field of protons and photons at saturation, having reached threshold, during its differential interaction on the atoms of the membranes and protein complexes of the mitochondrial intermembrane trap. Its resolution would initially give cold solutions in the realm of quantum physics (pseudo-condensation, vortexes), and hot solutions in the thermo-hydro-dynamic realms of statistical physics in a highly integrated biological environment.

Finally, a more problematic ëtate remains to be considered, which would be "*more spread out in space-time*" by the formation of a drifting field with a divergent tendency whose discrete non-linear consequences would be hidden by the sum of the superimposed states, which we have just highlighted and which are discussed in the conclusion. In particular, certain aspects that would be more or less discernible, called bio-photons with luminescent effects, or electroporation processes that would manifest themselves as a function of frequency ranges, if not deleterious chain reactions.

Low biological energies
and integration level
units
in theory
complex evolutionary processes.

In 1970, three researchers, Crowley, Neumann and Zimmermann, observed that electric fields increased the permëabilitë of the cell membrane by modifying its structure through ëlectroporation.

This is how pulsed signals, such as radar waves, will have biomedical applications in electrochemotherapy, a treatment carried out using a pulsed electric field with a high amplitude of the order of a kilo-volt/cm and a very short duration, applied locally in a few micro-seconds. This electromagnetic configuration modifies the permeability of the membrane and facilitates the penetration of an appropriate treatment into the cell.

From these experiments involving intracellular manipulation using pulsed fields in the nanosecond range, we have learned that the cell's response is not homogeneous from an electromagnetic point of view, due to its composition and the complex interactions that take place within it. The plasma membrane contributes to cellular exchanges via proteins that cross its lipid bilayer, either during passive diffusion of ions by osmosis[29] or during active transport via specific spë channels. The average resting potential of the plasma membrane is around seventy millivolts.

Comparatively, the mitochondrion stores Ca^{2+} and H ions in the matrix maintained at low pH, this concentration of H^+ ions is more ëкуë in the intermembrane space, as a result the inner membrane has a resting potential of less than one hundred and fifty millivolts correlative to the demands of intramembrane electron flow. Expërimentë on a cell or tissues (Thi Dan Thao Vu, 2012) a ë^й^^ pu^ field induces a dëplacement of charges on the external and internal faces of the membranes. This potential adds to that of the resting potential to reach a threshold of about one volt, causing a diiferential rearrangement of the atoms of each membrane and its protëine complexes, increasing its permëabilitë by modifying its conductivity within certain limits. This potential is reversible up to a critical value beyond which the cell becomes aerated, ultimately compromising its viability. These experiments demonstrate that for fields of amplitude greater than one millivolt/m and lasting from nanoseconds to a few hundred seconds, intracellular structures react to these brief pulses. Depending on an amplitude of less than a millivolt/m and the duration of a ёкание pue field smaller than a millisecond, the effective potential must be reconsidered, which is accompanied by a dërive towards more complex and discrete physiological processes, naturally structuring and/or dëstructuring, which manifest themselves from the simple creation of pores to apoptosis[30].

I . electroporation is thought to be an instantaneous process of impermëabilisation which tends to extend over time with electric fields of the order of one nanosecond, affecting the membranes of organelles such as mitochondria, which show a sensitivity to these waves estimated at four nanoseconds and ten millivolts/m. As a result, the pores created in the

---

29 Cutaneous respiration by osmosis is described in "L'Euprocte des PyrciKes".
30 Apoptosis is the programmed death of cells.

internal membrane release cytochromes C, activating caspases and triggering chain reactions that lead to apoptosis and programmed death, depending on the type of cell.

Numerous bioenergetics experiments have demonstrated the existence of electrochemical pulses in the mitochondria of plant and animal cells, probably pre-existing in prokaryotes. Several hypotheses have been formulated to justify these pulses:

- The regulatory activity linked to ATP synthase, which we have already mentioned.
- Mitochondrial anion transport.
- The flow of calcium and sodium.

On the other hand, "*underlying processes*" are cited (Markus Schwarzlander et al, 2012) that would involve the membrane potential and the pH gradient, whose electrogenic ion flow would be correlated to the coupling of electrons and protons in the inner membrane. In plants, an influx of Ca^{2+} would cause a pulsation by a fall in this potential, which would recover in a very short time. This decoupling could be used to counter mitochondrial dysfunction and slow the production of ROS. This phenomenon, which has been tested in induced stress states, has been studied in bacteria and animals.

Following Alexander Gurwitsch's research, the biophysicist Fritz-Albert Poop (1938-2018) established the existence of light impulses by ëvoking ultra-weak cellular radiation which, according to V. Mikhail Inyuschin (1975), would come from a field with regulatory properties encompassing the organism, like a bio-plasma. This research into bio-photons will often be contradicted and dëcrëdibilisëes by pseudo-scientific publications which will seize upon this so-called '*subtle*' energy by using the term quantum inconsiderately without argument, at the risk of losing any truly scientific meaning. How can we avoid navigating these murky waters?

34. A naval officer takes stock,
(Ecole Navale
statue).

Let's take stock of our naturalistic explorations of low biological ënergies. We have previously described the ability of a pulsed proton field to coordinate the cell's activity when it reaches saturation and threshold, suggesting the possibility of a discrete spreading of its quantum effects to divergent derived waves, a priori, with nonlinear effects that are unpredictable over a long period of time.

What happens if a disturbance in the flow of electrons causes a shortage or an excess in the proton field, and we deviate from the normal values for cellular activity?

According to the experimental observations we have just cited, the "*firing window*" *of* the proton and photon pulse field at saturation would be of the order of a few nanoseconds and about ten millivolts/m in amplitude. This value would appear to be too low to cause the heterogeneous and differential atomic rearrangement of the cell's proteins and membranes at threshold, ensuring their permeabilisation during the short phase of cellular activity. What's more, this coordinated process, which would confirm the proposal made by the Spanish biochemist Teresa Cordon, would manifest itself at different scales of resolution of integration level units - proteins, cells, animals - over a relatively long period of time. These perturbations of strictly quantum origin would be conditioned by random interferences that are difficult to predict at these different superimposed levels.

It emerges that the frequency and duration of the pulse field once it reaches saturation and threshold are important criteria during micro- and nanoporation, likely to modify the incident permeability of the inner and outer membranes of the mitochondria, and those of the cell: - At low frequencies, the wave would be non-penetrating, not reaching the inner membrane and its protein complexes sufficiently, although its local action on ion transport, proton translocation and biochemical displacements remains to be considered.

- At high frequencies, biological effects would be observed for a short-duration pulse limited to a firing window sufficiently effective to trigger a probable coordination of cellular activity to power the long mëtabolic phase, if only to ensure the maintenance of vital cohërence, with spëcific ëmergences at each integration level unit.

- However, the discreet underlying intermediate frequencies tend to provoke non-linear chain reactions, either naturally destructive or induced by external agents (radiation, disruptive chemicals, various stresses, etc.) with more or less discernible consequences (biophotons, luminescence) in pathological conditions close to lethality. Other physical criteria relating to the characteristics of the membrane and its components, the permittivity of the membrane and its properties, are also taken into account.[31] and membrane thickness should be included in the models.

In the previous chapter, using the non-linear equation of Gross and Pitaevskii, we demonstrated a divergent resultant wave that normally covers these low and high frequency ranges, but also incidentally, as a result of strong interactions, generates the induction of intermediate, more deleterious frequencies. Initially, these waves are characterised by a narrow peak and their scattering lengths become very much greater than zero, due to the effect of the couplings, so their probability density would be an indicator of a fleeting pseudocondensate. What we consider to be a specifically quantum cold solution with possible vortex formation, these vortices would extend into hot solutions in the domain of structuring and/or destructuring thermo-hydro-dynamic fluctuations-dissipations, under genetic and epigenetic control.

In our model, applied to units of integration level, protein, cell, animal, in the context of complex processes in biological evolution and transformations of civilisation, we propose :

That the threshold quantum field, pulsed by the charge of protons and photons that have reached saturation in the intermembrane space, acts differently on the membrane and the protein complexes. The differential susceptibility of the atoms of the inner membrane and the protein complexes to this threshold-pulsed field presents singularities that I have conveniently called 'cold solutions' to distinguish indistinguishable quantum processes from the more discernible and observable thermo-hydro-dynamic processes known as 'hot solutions'.

For the inner membrane of the mitochondrion, assimilated to a nematic liquid crystal in a

---

31 *The permittivity corresponds to the wave propagation speed in a more or less dispersive medium, normal or abnormal. If the high dispersion is abnormal at narrow band, the atom is condensed.*

layer associated with protein complexes, similar to a cholesteric liquid crystal, we have demonstrated in the models described in the previous chapters :

- A "cold" quantum solution, in which the positive composite wave of protons and photons at saturation, pulsed at threshold, would tend towards an ephemeral state of pseudo-Bose-Einstein condensate accompanied by a divergence of its resulting derivative waves, at several levels of frequency and energy resolution. This range would be biologically effective in a short time at low temperature, giving rise to high entropy at low energy, provoking genetically and epigenetically controllable chain reactions accompanied by a decrease in entropy and an increase in energy as vital coherence is achieved.

This would become uncontrollable and destructive through an increase in entropy and energy, in excess, during chain reactions that would incidentally take place over a long period of time.

- these complex coordinated biochemical processes would result in a range of 'hot' thermo-hydro-dynamic solutions through fluctuations and dissipations, sufficiently structuring and/or destructive for each unit of integration level, with a fundamentally random component, resulting from the interferences of quantum couplings in the cellular milieu. These processes would take place within the limits of the vital coherences required to maintain the animal's homeostasis and the species' ecological valence during its contingent evolution, which we evaluate in terms of its stochastic adaptability, which corresponds to the ratio of its stabilising variability subject to these fundamental aleas, to the randomness of external selective constraints, both natural and anthropogenic.

Since *these* two solutions interact with each other, it would be advisable to define the domain of crШсИё intermediary frequencies likely to disrupt the coordination of cellular activity in the mitochondria. This is clearly outside my exploratory capacity as a naturalist, and having almost reached the end of this essay, I can only suggest that we pursue this important line of research.

If the classical thermodynamic resolution is solved by the production of a ёlёvation of tempёrature, the quantum field resulting after interaction is considered to be of the "*divergent wave packet*" type. Fluctuations are normally dissipated in the more regular cycle of a free ёergy that spreads out over time by increasing, ensuring the caloric stabile of the body, whether hётёrothermic or homёothermic, at low biological ёne^e, Ipirticularly during hibernation. The consequences of the strongly non-linear ateatory effects of this resulting quantum field, very divergent in excёs, on entropy and chemical potential at very low tempёrature should be appreciated. In particular, the rearrangement of the atoms and modules that open the pores and optimise the sites of electron transfer and highly unusual proton translocations during the short phase of cellular activity that '*feeds*' its long metabolic phase.

According to our proposals dёveloppёes in the preceding chapters, at low tempёrature when the particles of cold atoms have a wave behaviour which is singularized by their de, de Broglie wavelength which decreases, this atomic dens^ of bosons would form a pseudo Bose-Eeinstein condensate, which possёderait :

-	A long-range cohёrence, this singular state of bio-plasma, a coordination factor yet to be proven, and its consёquences to be assessed at several levels.

-	A superfluid^, with vortex formation, in the absence of viscos^ it would form rotations by the nucteation of vortices with dissipative thermo-hydro-dynamic extensions. These are potential germs whose impact on the biological environment (stem cells, bacteria, viruses) and vital cohёrence during development, mёtamorphosis and the ^ё^^ incidental to an organ should be defined.

This ëtat is observable in weakly interacting gases (Mimoum, 2010) described by Gross-Pitaevskii and Bogoliubov mean field theory, which is a concept of macroscopic waves in a situation of collective excitations. Expërimentally, to act on this ëtate, it would be possible to proceedëto a tuning of these interactions by configuring the mësoscopic piëgeage associated with the anisotropic confinements of highly excited H+ protons *and* virtual photons, a pre-existing state in the mitochondrial intermembrane space.

This experimental resolution appears to be achievable by spinor treatment of the degrees of freedom linked to the proton spin[32] of the protons, either by acting on the flow of electrons, or by perturbing the proton field (pH, charge distribution, effective cross-section) or by acting on the trap atoms in the cell's mitochondria. A delicate disturbance to achieve on macroscopic superposition states with maximum correlations, for these ions (protons, electrons, free radicals) trapped at the frontier of the quantum and thermo-hydro-dynamic domains of integration level units (proteins, cell, animal).

While the repulsive interactions of the highly excited $H+*$ protons tend to expel the excitations from the trap, collisions between atoms rapidly modify the wave function at the origin of the pseudo-condensate, which is verified by the Gross-Pitaevskii equation, taking into account the more or less divergent derivative interactions. The trap would have to be made very anisotropic, for example for a confinement frequency of 3 khz, we would have an interaction term of 10^{-3} 8 kg m^3 s^{-2} and excited states of the atoms by their interactions and the temperature, outside the condensate.

In addition, we can see that the variations in energy states are associated with those in entropy, and note the notable influence of reversible ferromagnetic interactions:

For energy,

- With low energy, a single bound state, fairly high entropy, especially at very low temperatures.

- At high energies, critical phase shifts occur, and their functional impact needs to be demonstrated at very low and then at high temperatures.

For ferromagnetism, if we consider the eminent role of the iron oscillators included in the protein complexes of the respiratory chain.

- Ferromagnetic interactions maximise the total spin by aligning the spins.

- Antiferromagnetic interactions minimise spin, so the total spin state is low.

These interactions create, on the one hand, correlations between the atoms of the condensate and, on the other hand, excitations in indistinguishable non-classical states.[33]This would confirm our model of a complex evolutionary process which, for values of the coefficient of adaptability of less than two subject to alea, proves to be structuring and/or destructuring, including in the intermittent phases of strong non-linearity, which appear to be stable.

For this quantum solution that concerns us, according to the Bose-Einstein Condensate theory, the super fluiditë of dëgënërës gases has been dëmontrëed by the observation of vortices in rotating condensates. In mitochondria, the highly anisotropic confinement by the constituent atoms of the intermembrane space, constitutes a multiple Gaussian piëge with several ëtats of liquid pseudocrystals (membranes, protëines) for super-excited protons with an initially repulsive, then attractive, dominant when the weak field changes to very strong. At saturation,

---

32 Angular momentum.
33 Faintly discernible in a state of stress or close to letalite.

this complex field is likely to produce a very short, high-frequency pulse wave, which becomes more attractive, at threshold, when it interacts with the trapped atoms in the mitochondrial and cellular environment. The question is whether, at this level of saturation, there is a probability of a tendency towards the formation of Bose Einstein condensates at low critical temperature, where in all probability some protons and photons would be at the same wave function in one or more Gaussian traps.

In these magnetostatic traps (Tannoudji,1997/1998), under the effect of magnetic field gradients and magneto-caloric cooling after elastic collision between the trapped protons, can some of them acquire sufficient excitation energy to escape the trap by giving a *"cold, non-linear divergent"* quantum solution with a classical or alternating thermo-hydro-dynamic effect?

At the expense of the others, which could acquire a lower temperature after rethermalisation, during the "cold" quantum solution of indistinguishable pseudo-condensation.

In the latter case, can we actually establish a link between the microscopic properties of quantum solutions "cooled" by the formation of a trap potential well ($< 0°C$) at the mesoscopic level, and those macroscopic with the manifestation of a ëventual incidental condensation at low biological energy?

With regard to the quantum solution, which would tend towards an extremely divergent non-linear state, we need to recall the modalities of classical chaos and the transition to quantum chaos, according to Hadamard, Henri Poincare, Birkhoff and Anosov. According to David Ruelle, trajectories are highly sensitive to initial conditions and reveal themselves to be unpredictable, characterised by emergent random properties. To make the transition from classical chaos to quantum chaos, researchers are studying the dynamics of quantum waves to understand how they evolve, in order to estimate the consequences for their structure and assess their spatial distribution as standing waves. In the context of symplectic geometry, when studying the behaviour of a system with stable or unstable non-linear trajectories during its evolution, we use the notion of complex interleaving. Hamilton's equation, expressed geometrically independently of coordinates, suggests that wave mechanics is indeed *'hiding'* behind classical mechanics, and the classical/quantum correspondence is called micro-local analysis. A distribution of points initially concentrated outside stable equilibrium disperses in an unstable direction and converges towards the equilibrium distribution in a non-linear mixture. This transition state is adapted to Anosov flows.[34] flows for a semi-classical analysis. This quantum chaos would be a quantum dynamics defined by the Schrodinger equation whose Hamiltonian is such that the classical flow would be a non-linear solution closer to the solutions of the Gross-Pitaevskii equation, which we have used.

For the field of excited protons interacting with the heterogeneous atoms of membranes and specific to protein complexes, we referred to the repulsions of the ultra-fine structure levels at the complex interferences, initially intertwined, excessively mixed during the interaction. Paul Ehrenfest's *'long time'* quantum resolution of fluctuations around the mean would be subject to recurrence by a return to the initial instant, which would accentuate the non-linearity, increase the divergent process and the chain reactions normally controlled by genetic and epigenetic processes, or become uncontrollable, which raises a number of questions for us.

So for a periodically open system like the mitochondrion, if we assume that the threshold quantum field pulse coordinates the cell's short phase of activity. It should be remembered

---

34 A hyperbolic Anosov system would have extremely chaotic dynamics.

that particles (electrons, protons, ions) can escape to infinity during quantum diffusion with discrete resonances such as the non-linear quantum solution with singular local maxima at complex values that would influence the dynamics of cell activity over a long time. For example, by surreptitiously disrupting the classical thermodynamic resolution through deleterious oxidative chain reactions (free radicals, creation of pairs, etc.) capable of becoming naturally pathological, by altering gene regulation, which is highly dependent on low biological energies.

35. Cerf ëlaphe, lumiëre and sexual reproduction.
Biological structures are not static; vital cohesion depends on dynamic internal flows and exchanges with the environment, on matter (wave and corpuscle), both energy and information, an unstable out-of-balance state that evolves differently in space and time, causing :
- An inevitable degradation, such as ageing, pathologies and death.
- Coordinated and regulated rhythmic activities that momentarily guarantee vital coherence, reproduction and adaptability, during species revolution.
- The rates of growth and chain reaction are gënëtically controlled as each organism develops, metamorphoses and regenerates.
This has led us to define the initial conditions and limits of adaptability of a species, during its biological evolution, which is fundamentally stochastic, in its dual quantum aspect, random and indiscernible, now classically discernible, ultimately structuring and/or destructuring. These potential variations are confronted with the random circumstances of natural selective constraints, and those now generated in excess by our predominant social and economic activities, which disrupt the parameters of the adaptability relationship.
This naturalistic essay on low-energy biology brings us within reach of particle accelerators that are not very demanding in terms of energy, but of a bewildering complexity that combines the resolution levels of quantum physics with those of classical physics applied to the complex processes of biological revolution and the transformations of civilisation. After my experiences as a naval officer and my travels around the world, I made numerous naturalistic observations to further develop the theory of biological revolution and the transformations of civilisation by the reversal effect (Tort, 1983). In writing the four volumes

of the naturalist suite, this essay was an essential exploration of a major step towards a better understanding of these complex evolutionary processes.

In contrast to the low biological energy levels, our civilisation uses considerable energy to satisfy its most vital needs and to satisfy unconsidered desires, a subject that will be dealt with in *"L'Homme desarticule"*. This essay is intended to be a forward-looking assessment, a necessary reflection before writing the *'Seventh String'*, which will be the ultimate synthesis of my research, a kind of symphony, necessarily unfinished...

36. *Seven-string viola da gamba*

The Greenland Shark

Among the animals that live at low tempĕrature, the Greenland shark, *Somniosus microcephalus*, lives in the icy depths of Arctic waters, and is characterised by a long lifespan, measuring 7.3 metres in length and weighing 1,200 kg. Its cylindrical body is covered by skin that is a mixture of brown, shades of grey and black, covered with sharp skin denticles and placoid scales. This shark lives between the surface and 2,200 metres, reaching a depth of 2,774 m where the pressure is 277 bars. It swims very slowly, almost motionless, and is viviparous, with the offspring developing and hatching inside the female, without a placenta.

Its physiology and metabolism are adapted to the depth, salinity and low temperatures of the Arctic depths. As with the skin cells of the Pyrenean mudskipper (Calotriton asper asper), the osmotic and respiratory functions of those of the Greenland shark raised questions in my mind, in particular the process of producing a toxic product that gives a strong flavour to the shark meat so highly prized by Greenlanders.

In addition to the seawater it ingests and discharges through its gills as it moves slowly around the body, seawater and solutes pass through the body by osmosis through the skin cells, inversely irrigating the tissues and organs. Proteins and trimethylamine oxide, produced by the metabolic decomposition of protëines and amino acids under extreme conditions of temperature and pressure, contribute to this direct osmosis with seawater, but do not cross the selective membranes involved in salt balance.

Usually in fish, when salt and other solutes penetrate the tissues, water is expelled from the body. As the concentration of salt in marine fish is lower than in seawater, fish must continually absorb water through their gills to excrete the excess salt.

The concentration of salt in the Greenland shark is also lower than in their environment, but they manage osmosis differently. In order to maintain a stable quantity of water in its body, the Greenland shark retains a high concentration of urea in its blood, which compensates for the lower concentration of salt localized in its skin cells. However, because a high level of toxic urea would damage the body by destabilising proteins, the Greenland shark must also retain an even higher level of trimethylamine oxide to counteract the effects of the urea. When trimethylamine oxide and urea are combined with the salt in the Greenland shark's tissues, the osmotic pressure of its body fluids becomes higher than that of its environment, enabling it to withstand the pressure of the deep sea.

As a result, the Greenland shark is dirtier than sea water. Unlike bony fish, which must constantly and actively ingest water to replace the water lost through osmosis, the Greenland shark does not need to expend energy to maintain the level of water required to keep it alive.

As well as contributing to the shark's osmotic pressure, 1 trimethylamine oxide and urea act as a natural antifreeze that stabilises enzymes and protëines in the Greenland shark's tissues. When the shark goes through extreme conditions of low tempĕratures and extreme depths, they prevent the formation of ice crystals that would perforate cell membranes, leading to the loss of cell contents, organ destruction, and death.

When Greenland shark meat is consumed, the digestive process transforms trimëthylamine oxide into trimëthylamine, a substance with a strong odour of ammonia or rotten fish. As well as causing intestinal pain, trimëthylamine has a neurological effect akin to excessive alcohol consumption. In extreme cases, death can follow when too much flesh has been consumëe.

The Greenland shark is nevertheless considered a favourite dish in Iceland, where Hakarl is eaten. Beyond this culinary aspect, it is a subject of study on low biological energies and an invitation to explore the poles.

70

Bibliography

- Albert Einstein, *La relativite*, la petite bibliotheque Payot, 1963.
- Andrianaivomananjaona Tiona, *Mecanismes de regulation de l'ATP Synthase mitochondriale de S. cerevisiae par son peptide endogene IF1 et étude de l'oligomerisation d'IF1 de S. cerevisae, thesis univer site Paris XI, 2011.*
- Arneado Alain, Argoul Frangoise, Bacry Emmanuel, Elezgaray Juan, Muzy Jean-Francois, *Ondelettes, multifractales et turbulences*, Diderot, 1995.
- Bernal J.D., Haldane J.B.S., Pirie N.W., Pringle J.W.S., *A discussion of the origin of life*, Rationalist Union, 1955.
- Brackett S. Frederick, *The presents state of physics*, The american association for the advancement of science, 1954.
- Cassë Michel, *Du vide a la creation*, Editions Odile Jacob, 1995.
- Cohen-Tannoudji Claude, *Etude de la condensation de Bose- Einstein des gaz atomiques ultra-froids, cours 1997/1998/1999. diatomiques ultra-froids*, cours du College de France 1997-1998.
- Cohen-Tannoudji Claude and Spiro Michel, *Le boson et le chapeau mexicain*, Gallimard, 2013.
- Cottet-Rousselle Cecile, *Mesure par microscopie confocale du metabolisme mitochondrial et du niveau energetique cellulaire au cours d'épisodes de carences en substrats et/ou en oxygene*, PhD thesis in cellular and molecular biology and health sciences, Ecole Pratique des Hautes Etudes, 2016.
- Cox Brian and Forshaw, *The Quantum Universe,* Dunod, 2013.
- Cunchillos Chomin, *Les voies de l'emergence*, Belin, 2014.
- De Duve Christian, *Singularites*, Odile Jacob, 2011.
- Feynman Richard, *The Nature of Physics*, Seuil, 1980.
- Feynman Richard, *Lumiere et matiere*, InterEditions, 1987.
- Popp Fritz-A, *Biologie de la lumiere*, Editions Marco Pietteur, 1989.
- Froger Jean-Francois and Lutz Robert, *Fondements logiques de la physique*, DësIris, 2007.
- Glass-Maujean Michele, *Etude par la methode des anticrosements, des niveaux n=3 et 4 de l'atome d'hydrogene excite forme par dissociation de la molecule H2*, These de doctorat d'Etat es sciences physiques, Universite de Paris, 1974.
- Herzberg Gerhard, *Atomic spectra and atomic structure*, Dover publications, 1944.
- Jackson L.C., Low temperature physics, Methuen,1950.
- Landau Lev and Lifchitz E, *Physique statistique*, Ellipses, 1994.
- Markus Schwarzlander, David C. Logan, Iain G. Johnston, Nick S. Jones, Andreas J. Meyer, Mark D. Fricker, Lee J. Sweetlove, *Membrane potential pulsation in individual mitochondria: a stress-induced mechanism to regulate bioenergetics in Arabidopsis*, the plant cell, 2012.
- Mayr Ernest, *Populations, especes et evolution*, Hermann, 1984.
- Mimoun Emmanuel, *Sodium Bose-Einstein condensate in a mesoscopic trap*, Université Paris 6, 2010.
- Monod Jacques, *Le hasard et la necessite*, editions du Seuil, 1970.
- Naisse J.P., *Effets de polarisation du vide dans la diffusion proton-proton en onde S*, Bulletin academie royale de Belgique, 1972 (pages 872/873).

- Nonnenmacher Stephane, *quelques aspects du chaos quantique*, thesis of 5 June 3009, Université Paris Sud, faculte des sciences d'Orsay.
- Prigogine I. and Glansdorff P., *Structure, stabilite et fluctuations,* Masson, 1971.
- Ruelle David, *Hasard et Chaos*, Odile Jacob, 1991.
- Rush J.H., *L 'origine de la vie*, petite biblotheque Payot, 1969.
-Schrodinger Erwin, *Qu'est ce que la vie*, Christian bourgeois, 1986.
- Schrodinger Erwin, *What is life?* 1986, Christian Bourgeois Editor.
- Schrodinger Erwin, *Physique quantique et representation du monde*, Seuil, 1992.
- Tolansky S., *Hyperfine structure in line spectra and nuclear spin*, Methuen,1953.
- Tort Patrick, *Pour Darwin*, puf, 1997.

particularly around zero of the axes, a zone of cold and hot solutions dëtaillëe in the graph prëcëdent.

33. Simulation using Gross and Pitaevskii liquidation.
34. A naval officer takes stock (Statue of the Ecole Navale).
35. Cerf ëlaphe, lumiëre and sexual reproduction.
36. Seven-string viola da gamba.
37. Lastiëre cover, author Claude Rouquette.

LOW ENERGY
ORGANIC

As a complement to the four volumes of the naturalist suite, this essay on low biological energies was needed to explore the transitions between quantum mechanics and statistical physics at work in the integration level units of living organisms, proteins, cells and animals. From questioning to questioning about the coordination of cellular activity, marine historian and naturalist Claude ROUQUETTE embarks on a voyage of discovery, probing the depths of low energies in order to take stock of the complex processes of biological evolution, intimately associated with the transformations of civilisation.

Former naval officer,
specialising in safety and the environment,
Claude ROUQUETTE is conducting
fundamental and applied research into complex evolutionary
processes. As an
author and lecturer, he is responsible for
promoting long-term scientific

Charles Darwin International, directed
by Professor Patrick Tort.

I want morebooks!

Buy your books fast and straightforward online - at one of world's fastest growing online book stores! Environmentally sound due to Print-on-Demand technologies.

Buy your books online at
www.morebooks.shop

Kaufen Sie Ihre Bücher schnell und unkompliziert online – auf einer der am schnellsten wachsenden Buchhandelsplattformen weltweit! Dank Print-On-Demand umwelt- und ressourcenschonend produziert.

Bücher schneller online kaufen
www.morebooks.shop

info@omniscriptum.com
www.omniscriptum.com

Printed by Books on Demand GmbH, Norderstedt / Germany